Contemporary Trends and Issues in Science Education

Volume 53

The book series Contemporary Trends and Issues in Science Education provides a forum for innovative trends and issues impacting science education. Scholarship that focuses on advancing new visions, understanding, and is at the forefront of the field is found in this series. Authoritative works based on empirical research and/or conceptual theory from disciplines including historical, philosophical, psychological and sociological traditions are represented here. Our goal is to advance the field of science education by testing and pushing the prevailing sociocultural norms about teaching, learning, research and policy. Book proposals for this series may be submitted to the Publishing Editor: Claudia Acuna E-mail: Claudia.Acuna@springer.com

Christine D. Tippett · Todd M. Milford
Editors

Exploring Elementary Science Teaching and Learning in Canada

 Springer

Editors
Christine D. Tippett
University of Ottawa
Ottawa, ON, Canada

Todd M. Milford
University of Victoria
Victoria, BC, Canada

ISSN 1878-0482   ISSN 1878-0784 (electronic)
Contemporary Trends and Issues in Science Education
ISBN 978-3-031-23935-9   ISBN 978-3-031-23936-6 (eBook)
https://doi.org/10.1007/978-3-031-23936-6

This Springer imprint is published by the registered company Springer Nature Switzerland AG
The registered company address is: Gewerbestrasse 11, 6330 Cham, Switzerland

Reviewers

Dana Bell, University of Victoria
David Blades, University of Victoria
Lynn Fels, Simon Fraser University
Nicole Ferguson, Université de Moncton
Rob Kiddell, University of Victoria
Michael W. Link, University of Winnipeg
Andrew MacLean, University of Victoria
Barbara McMillan, University of Manitoba
Wendy Nielson, University of Wollongong
Lilian Pozzer, University of Manitoba
Latika Raisinghani, University of Regina
Carol Rees, Thompson River University
Giuliano Reis, University of Ottawa
Darron Kelly, Memorial University of Newfoundland
Sharon Pelech, University of Lethbridge
Jrène Rahm, Université de Montréal
Natalie Sadowski, University of Ottawa
Astrid Steele, Nipissing University
Poh Tan, Simon Fraser University
Grant Williams, St. Thomas University
Roxana Yanez Gonzalez, University of Ottawa
Larry Yore, University of Victoria

Acknowledgements

It has taken much effort to make this publication a reality.
We would like to thank:

- Our contributing authors, who shared personal stories of their research and had faith in us as editors
- Our reviewers, who willingly took the time to provide feedback that would improve the chapters
- Our editor, Claudia Acuna, who was so patient with us as we tried our best to keep the publication moving forward and meet deadlines in the middle of a global pandemic
- Our copyeditor, Shari Yore, whose expertise we relied on yet again—but all errors are ours alone
- Our mentor, Larry Yore, whose generosity and scholarly personality we aspire to emulate
- Our friends, family, and loved ones, who continue to support us

Christine D. Tippett
Todd M. Milford

Contents

Editors and Contributors

About the Editors

Christine D. Tippett BASc (University of British Columbia), BEd (University of Victoria), MA (University of Victoria), PhD (University of Victoria), is an associate professor of science education in the Faculty of Education at the University of Ottawa. She was an engineer before she obtained her teaching degree, which influences her ways of thinking about science education. Her research interests include visual representations, science education for all students, and professional development for science educators (preservice, inservice, and informal). Current projects focus on preservice science teachers' images of engineers, early childhood STEM education, and assessment of representational competence.

Todd M. Milford BSc (University of Victoria), BEd (University of Victoria), DipSpecEd (University of British Columbia), MEd (University of Victoria), PhD (University of Victoria), is an associate professor and chair in the department of Curriculum and Instruction at the University of Victoria. Prior to this he was a lecturer in the Art, Law, and Education Group at Griffith University in Brisbane Australia. He has science and special education classroom teaching experience as well as in the online environment. He has been teaching at the postsecondary level since 2005 primarily in the areas of science education, mathematics education, and classroom assessment. His research has been and continues to be varied; however, the constant theme is using data and data analysis to help teachers and students in the classroom.

Contributors

Saiqa Azam Memorial University of Newfoundland, Faculty of Education, St. John's, NL, Canada

G. Michael Bowen Faculty of Education, Mount Saint Vincent University, Halifax, NS, Canada

Xavier Fazio Faculty of Education, Department of Educational Studies, Brock University, St. Catharines, ON, Canada

Lynn Fels Simon Fraser University, Burnaby, BC, Canada

Karen Goodnough Memorial University of Newfoundland, Faculty of Education, St. John's, NL, Canada

Kamini Jaipal-Jamani Faculty of Education, Department of Educational Studies, Brock University, St. Catharines, ON, Canada

Jesse Jewell Yukon Wild School, Whitehorse, Yukon, Canada

Darron Kelly Memorial University, St. John's, NL, Canada

Eva Knoll Département de Mathématiques, Université du Québec À Montréal, Montréal, Québec, Canada

Michael W. Link University of Winnipeg, Winnipeg, MB, Canada

Karen Meyer University of British Columbia, Vancouver, BC, Canada

Todd M. Milford Faculty of Education, University of Victoria, STN CSC, Victoria, BC, Canada

Tim Molnar College of Education, University of Saskatchewan, Saskatoon, SK, Canada

Sharon Pelech University of Lethbridge, Lethbridge, AB, Canada

Carol A. B. Rees Thompson Rivers University, Kamloops, BC, Canada

Dawn L. Sutherland University of Manitoba, Winnipeg, MB, Canada

Poh Tan Faculty of Education, Simon Fraser University, Burnaby, BC, Canada

Christine D. Tippett University of Ottawa, Ottawa, ON, Canada

Patrick Wells Memorial University of Newfoundland, Faculty of Education, St. John's, NL, Canada

Amy M. Willison Independent Consultant, Halifax, NS, Canada

Chapter 1
Providing a Space for Canadian Science Education Research

Christine D. Tippett and Todd M. Milford

1.1   Introduction

Canada is a multicultural nation with a diverse geography and a correspondingly complex history. Its governance is typically federal, provincial/territorial, or municipal; education—curriculum development, instruction, and student assessment—is a provincial/territorial responsibility (for an extended discussion, see Milford & Tippett, 2019). When it comes to science education, Canadian students have been consistently identified as top performers in international assessments (OECD, 2019; Statistics Canada & Council of Ministers of Education, Canada, 2019). This strong performance can be attributed to factors at the macro, meso, and micro levels. At the macro level, Canada's education system can be compared to other nations' systems; for example, the Programme for International Student Assessment reports on multiple predictors can be used to compare aspects of different countries' educational success. At the meso level, Canada's educational system can be examined as a whole, which we did in an earlier edited book (Tippett & Milford, 2019). There we presented an overview of Canada's science education system and identified consistencies, commonalities, and distinctions across its 13 jurisdictions. At the micro level, specific aspects of Canada's education system, such as particular pedagogical approaches or highly contextualized research results, can be considered. With this edited book, we add to the micro-level literature on elementary science education research being conducted in Canada.

C. D. Tippett (✉)
University of Ottawa, 145 Jean-Jacques-Lussier, Ottawa, ON K1N 6N5, Canada
e-mail: ctippett@uottawa.ca

T. M. Milford
Faculty of Education, University of Victoria, STN CSC, PO Box 1700, Victoria, BC V8W 2Y2, Canada
e-mail: tmilford@uvic.ca

© The Author(s), under exclusive license to Springer Nature Switzerland AG 2023
C. D. Tippett and T. M. Milford (eds.), *Exploring Elementary Science Teaching and Learning in Canada*, Contemporary Trends and Issues in Science Education 53, https://doi.org/10.1007/978-3-031-23936-6_1

1.2 The Uniqueness of Canadian Science Education Research

The topics that are explored at the micro level contribute to the unique nature of Canadian science education research. To situate our national research against a global backdrop, we developed a master list of research topics that have been identified nationally and internationally. We began with the nine categories that were initially derived from the conference strands of the National Association for Research in Science Teaching (Tsai & Wen, 2005) and used in a content analysis of three international science education journals that reported on research topics appearing between 1998 and 2017 (Lin et al., 2018).

Then, to understand Canadian research interests, we turned to an analysis of the presentations made at the annual Science Education Research Group (SERG) events between 2013 and 2017 (Tippett et al., 2017). SERG is Canada's only national organization for science education researchers, making its event a key avenue for the dissemination of Canadian science education research results. The 10 most frequent SERG topics were teaching, STEM (science, technology, engineering, and mathematics), inquiry and other skills, preservice teachers, environmental issues, science content knowledge, student focused, identity and self-efficacy, professional development, and social justice and diversity (Tippett et al., 2017). We added these topics to our master list, aligning with the terminology of the international trends where possible.

The last contributions to the master list were the nine themes that we identified in an analysis of the chapters in our earlier edited book (Tippett et al., 2019): low priority of science, influences on curricula, Indigenous perspectives, inadequate science teacher education, language issues, assessment practices, locally developed courses, instructional approaches, and environmental considerations. Despite the focus of that book being education rather than research, we anticipated that its themes might help provide a fuller picture of distinctly Canadian research. Again, we aligned themes with the items already in the master list and removed any duplicate items. Finally, we alphabetized the master list and then compared all SERG research topics to that list, as shown in Table 1.1.

It is clear that Canadian researchers do explore science education topics that are common globally, but it is also clear that they examine topics outside the list of international trends. For example, the topics of assessment, environmental issues, identity and self-efficacy, Indigenous perspectives, inquiry and other skills, professional development, STEM and engineering, and student-focused research do not appear as international trends. Although we know that there *is* research being conducted globally about these topics, their prominence in Canadian science education research can be considered unique.

Despite its unique nature, however, there is no dedicated space for the dissemination of the results of science education research in Canada. Presently all avenues that exist are neither wholly Canadian nor science focused. The *Canadian Journal of Science, Mathematics and Technology Education* was originally established to

Table 1.1 Science education research topics, international and Canadian

Master list of topics	International trends [a]	Top 10 SERG topics [b]	Science education themes [c]	All SERG topics [b]
Assessment			☆	☆
Culture, society, & gender	☆	☆	☆	☆
Educational technology	☆			☆
Environmental issues		☆	☆	☆
Goals, policy, & curriculum	☆		☆	☆
Identity & self efficacy		☆		☆
Indigenous perspectives			☆	☆
Informal learning	☆			☆
Inquiry & other skills		☆		☆
Learning (conceptions)	☆	☆		☆
Learning (context)	☆			☆
Philosophy, history, & nature of science	☆			☆
Professional development		☆		☆
STEM & engineering		☆		☆
Student-focused		☆		☆
Teacher education	☆	☆	☆	☆
Teaching	☆	☆	☆	☆

[a] Lin et al., 2018.
[b] Tippett et al., 2017.
[c] Tippett et al., 2019.

provide "a Canadian voice" (Hodson et al., 2001, p. 5), but the journal is not solely Canadian nor exclusively science or research oriented. In a content analysis, Pegg et al. (2015) found that only 56% of the items published during the journal's first 14 years were from authors with Canadian affiliations. Further, they reported that only 192 (of more than 350) articles, papers, and viewpoints were related to science education and that those items included 37 interdisciplinary items or items with "some connection to science" (p. 379).

1.3 About the Book

With this book we are creating that much needed space for uniquely Canadian science education research results. Our intent was to highlight the breadth of the research being done across the country. We also hoped to expand upon some of the themes from *Science Education in Canada* (Tippett & Milford, 2019) as we explored the particular and diverse nature of science education research in Canada. We put out a call for chapter proposals and received more than 20 responses, which led to our decision to publish two books: one elementary and one secondary/tertiary. The chapters in this elementary book mainly span pre-Kindergarten (pre-K) to Grade 7, with one chapter addressing K to Grade 9.

The book includes 11 chapters that present research conducted across the country and with a wide range of geographic settings, grade levels, student or teacher foci, topics, and methods (see Table 1.2). There are six chapters focusing on teachers, one chapter focusing on students, and four chapters focusing on teachers and students together. Our contributing authors displayed an obvious interest in teacher education with six chapters touching on aspects of discourse patterns, inclusive education, dramatic approaches, nature of science, instructional practices, and engineering design. Three chapters focus on the natural environment as a learning context (e.g., supporting children's well-being, children's authentic interest in nature, place-conscious pedagogy); other chapters touch on science literacy, Indigenous perspectives, and aspects of STEM. Authors followed qualitative, quantitative, and mixed methods approaches; specific research methods include case studies, document analysis, and phenomenology. Data were collected via artifacts (e.g., student products, student and teacher documents), observations (e.g., video/audio/fieldnotes), surveys, and writing exercises (e.g., evaluations, discussion posts).

When we compared the contents of the 11 chapters in this book to the original themes highlighted in our first book, we found less overlap than we had predicted. The two consistencies—a low priority of science and influences on science curricula—were not reflected. Of the four commonalities—Indigenous perspectives, inadequate science teacher education, language issues, and assessment practices—only inadequate science teacher preparation was clearly addressed with six chapters emphasizing teacher education. Indigenous perspectives were featured in two chapters, but language issues and assessment were not featured at all. The three distinctions—locally developed courses, instructional approaches, and environmental considerations—were evident, with seven chapters examining specific pedagogical approaches and/or novel courses and three chapters highlighting environmental contexts for learning.

The international nature of the topics addressed by Canadian science education researchers in this book means that chapters will be of interest to a wide range of readers. Examples of topics of international interest include *co-teaching* (Chap. 2), *inclusivity* (Chap. 3), *science and the arts* (Chap. 4), *science and technology-mediated pedagogies* (Chaps. 5, 9, and 11), *Indigenous perspectives* (Chaps. 6 and 8), *teacher*

Table 1.2 Overview of chapters

Chapter	Geographic setting[a]	Grade level	Focus Student	Focus Teacher	Teacher education[b]	Topics	Method[c] (Data sources)
2	BC	1	☆	☆	Pro-D	Co-teaching, discourse patterns, science inquiry, Steps to Inquiry framework	MM (observations, student/teacher artifacts)
3	NL	2		☆	Pro-D	Teachers in Action, Universal Design for Learning	QUAL (interviews, student/teacher artifacts, observations)
4	BC	K–7		☆	Preservice	Drama, performative inquiry, problem solving	QUAL (video, reflections)
5	MB	elementary		☆	Preservice	Community of practice, engineering design, outreach	QUAL (interviews)
6	AB	K–9		☆	Preservice	Data literacy, Indigenous perspectives, science methods course	QUAL (observations, feedback)
7	ON	elementary		☆	Pro-D	Blended format (face-to-face and online), science teaching innovation	MM (interviews, online discussions, surveys)
8	BC	pre-K	☆	☆		Indigenous perspectives, Kānaka Maoli worldview, scientific literacy	QUAL (observations, student/teacher artifacts)
9	NS	K–3		☆		Bee-Bots, STEM, social media, teacher resources	MM (online teacher resources)
10	MB	1–2	☆	☆		Experiences in nature, capabilities for well-being	QUAL (interviews, observations)
11	YT	2	☆			Environment, authentic explorations, wearable technology	MM (interviews, GPS data)
12	AB	4	☆	☆		Place-conscious pedagogy, student curiosity	QUAL (interviews, student artifacts)

[a] AB = Alberta, BC = British Columbia, MB = Manitoba, NL = Newfoundland and Labrador, NS = Nova Scotia, ON = Ontario, YT = Yukon Territories.
[b] Pro-D = professional development, preservice = teacher preparation/science methods instruction.
[c] MM = mixed methods, QUAL = qualitative, QUAN = quantitative.

professional development (Chaps. 2, 3, and 7), and *elementary science and the environment* (Chaps. 10, 11, and 12). Although we emphasize these particular topics as common internationally, readers may want to refer to Table 1.2 to locate chapters with additional topics of interest.

1.4   How to Read This Book

There are several ways in which this book can be read. When deciding how to present the chapters, we thought about various groupings or orders and eventually landed upon teacher education—one of the themes from our first book—as a broad organizer for the first half and then more specific chapters in the second half. Like any good nonfiction compendium, this book does not need to be read in a particular order. Readers can access chapters in whatever order they prefer: following grade level, according to topic, or by teacher/student focus. The book can be read as a full collection from cover to cover or as a set of individual chapters in isolation. We leave it to individual readers to choose what best suits their needs.

This book can serve several functions for readers. First, it can function as a stand-alone publication that offers a snapshot of the diverse undertakings of Canada's science education researchers. The contributing authors have contextualized their work, allowing readers to make micro-level comparisons with their respective contexts. Second, this book can function in combination with its forthcoming sister publication, which will be an examination of secondary and postsecondary science education research in Canada. The two companion books will thus present the entire pre-K to postsecondary space. Considering the similarities and differences across elementary, secondary, and tertiary science education research in Canada should be enlightening for Canadian and international readers alike. Finally, it can function as an extension of *Science Education in Canada* (Tippett & Milford, 2019), allowing further exploration of several of the themes identified in that book.

References

Hodson, D., Hanna, G., & Désautels, J. (2001). Finally, a Canadian voice. *Canadian Journal of Science, Mathematics and Technology Education, 1*(1), 5–7. https://doi.org/10.1080/149261501 09556447
Lin, T.-J., Lin, T.-C., Potvin, P., & Tsai, C.-C. (2018). Research trends in science education from 2013 to 2017: A systematic content analysis of publications in selected journals. *International Journal of Science Education, 41*(3), 367–387. https://doi.org/10.1080/09500693.2018.1550274
Milford, T. M., & Tippett, C. D. (2019). Introduction: Setting the scene for a meso-level analysis of Canadian science education. In C. D. Tippett & T. M. Milford (Eds.), *Science education in Canada: Consistencies, commonalities, and distinctions* (pp. 1–12). Springer. https://doi.org/10.1007/978-3-030-06191-3_1
Organisation for Economic Co-operation and Development. (2019). *PISA 2018 results (Vol. 1): What students know and can do.* https://doi.org/10.1787/5f07c754-en

Pegg, J., Wiseman, D., & Brown, C. (2015). Conversations about science education: A retrospective of science education research in CJSMTE. *Canadian Journal of Science, Mathematics and Technology Education, 15*(4), 364–386. https://doi.org/10.1080/14926156.2015.1093202

Statistics Canada & Council of Ministers of Education, Canada. (2019). *Education indicators in Canada: An international perspective 2019.* https://www150.statcan.gc.ca/n1/en/pub/81-604-x/81-604-x2019001-eng.pdf?st=aYxAx7Ci

Tippett, C. D., & Milford, T. M. (Eds.). (2019). *Science education in Canada: Consistencies, commonalities, and distinctions.* Springer. https://doi.org/10.1007/978-3-030-06191-3

Tippett, C. D., Milford, T. M., & Yore, L. D. (2019). Epilogue: The current context of Canadian science education and issues for further consideration. In C. D. Tippett & T. M. Milford (Eds.), *Science education in Canada: Consistencies, commonalities, and distinctions* (pp. 311–337). Springer. https://doi.org/10.1007/978-3-030-06191-3_15

Tippett, C. D., Pegg, J., Wiseman, D., & Milford, T. M. (2017, August 21–25). *Canadian science education research: A comparison across dissemination outlets* [Paper presentation]. European Science Education Research Association.

Tsai, C.-C., & Wen, M. L. (2005). Research and trends in science education from 1998 to 2002: A content analysis of publications in selected journals. *International Journal of Science Education, 27*(1), 3–14. https://doi.org/10.1080/0950069042000243727

Christine D. Tippett BASc (University of British Columbia), BEd (University of Victoria), MA (University of Victoria), PhD (University of Victoria), is an associate professor of science education in the Faculty of Education at the University of Ottawa. She was an engineer before she obtained her teaching degree, which influences her ways of thinking about science education. Her research interests include visual representations, science education for all students, and professional development for science educators (preservice, inservice, and informal). Current projects focus on preservice science teachers' images of engineers, early childhood STEM education, and assessment of representational competence.

Todd M. Milford BSc (University of Victoria), BEd (University of Victoria), Dip SpecEd (University of British Columbia), MEd (University of Victoria), PhD (University of Victoria), is an associate professor and chair in the department of Curriculum and Instruction at the University of Victoria. Prior to this he was a lecturer in the Art, Law, and Education Group at Griffith University in Brisbane Australia. He has science and special education classroom teaching experience as well as in the online environment. He has been teaching at the postsecondary level since 2005 primarily in the areas of science education, mathematics education, and classroom assessment. His research has been and continues to be varied; however, the constant theme is using data and data analysis to help teachers and students in the classroom.

Chapter 2
Changes in Discourse Patterns During Scientific Inquiry: A Co-teaching Model for Teacher Professional Learning

Carol A. B. Rees

2.1 Introduction

One of the goals inherent in K–12 science education curricula across Canada and other parts of the world, especially the United States, Europe, and Australia, is to provide students with opportunities to engage in the practices of scientific inquiry, which include asking questions, planning and carrying out investigations, and analyzing and interpreting data (e.g., British Columbia Ministry of Education, 2016; Ontario Ministry of Education, 2007; Rocard et al., 2007; Tytler, 2007; United States National Research Council, 2013). Students need opportunities to use "the methods and procedures of science to investigate phenomena, test and develop understanding, solve problems and follow interests" (Hodson, 2014, pp. 2545–2546). In this kind of inquiry, students perform activities of scientific investigation; and they need opportunities to share ideas through dialogic discourse at all stages of the process (Lehesvuori et al., 2011).

The recommendation to include scientific inquiry extends to elementary curricula. Elementary teachers often have little science education background, which makes this curricular recommendation particularly difficult for them to achieve (Steele et al., 2013). Accordingly, various frameworks have been developed to support elementary teachers and their students in scientific inquiry. One such framework is the Steps to Inquiry Framework (SIF; Pardo & Parker, 2010); however, one issue with learning to use SIF is that teachers can find it difficult to transfer their new knowledge into the classroom. This chapter focuses on a co-teaching model for teacher professional learning that involved two professionals: an expert teacher with a science background

C. A. B. Rees (✉)
Thompson Rivers University, 805 Tru Way, Kamloops, BC V2C 0C8, Canada
e-mail: crees@tru.ca

© The Author(s), under exclusive license to Springer Nature Switzerland AG 2023
C. D. Tippett and T. M. Milford (eds.), *Exploring Elementary Science Teaching and Learning in Canada*, Contemporary Trends and Issues in Science Education 53,
https://doi.org/10.1007/978-3-031-23936-6_2

and experience using SIF, and a novice teacher who was learning to incorporate scientific inquiry into his Grade 1 classroom using SIF. We were particularly interested in analysing the discourse patterns that occurred during each phase of the co-teaching model. The research question addressed was: How did discourse patterns change in a Grade 1 science classroom throughout a co-teaching experience?

2.2   Literature Review

The literature review provides a brief history of scientific inquiry along with an introduction to SIF. The co-teaching model and its use for teacher professional learning in science education are described, and teacher–student and teacher–teacher discourse patterns are discussed.

2.2.1   Scientific Inquiry

The term *scientific inquiry* refers to the particular ways of observing, thinking, investigating, and validating that scientists use in their work (American Association for the Advancement of Science, 1993/2009). Scientific inquiry in the classroom begins with students developing their own questions then designing and conducting their own scientific investigations. Efforts to implement scientific inquiry have a long history in North America beginning with Dewey (1910), who introduced the idea that students need opportunities to engage with the practices of science and scientific thinking as well as opportunities to learn science as a subject matter. He later proposed that the questions students investigate need to relate to their own experiences (Dewey, 1938).

The focus on scientific inquiry in school science curricula gained prominence in North America by the 1960s (Schwab, 1960, 1962). This prominence continued through the 1970s as indicated by the National Science Teachers Association (1971) position paper on science education that recommended students have "an opportunity for investigative activities involving open inquiry" (p. 49). However, typical school science practical experiences were teacher-directed; for example, the teacher provided a question and a plan for students to follow to achieve a predetermined answer. Researchers began advocating for authentic scientific inquiry experiences for students that were more open and more akin to the practices of scientists (e.g., Hodson, 1996; Roth & Bowen, 1995).

Studies indicate that scientific inquiry approaches where students generate questions, design experiments, collect data, draw conclusions, and communicate findings—all of which emphasize students' active thinking and responsibility for learning—are associated with increased interest (Anderson, 2002; Kang & Keinonen, 2018; Minner et al., 2010), motivation (Tuan et al., 2005), and improved science learning as long as the inquiry is appropriately guided by teachers (Aditomo & Klieme, 2020; Furtak et al., 2012; Lazonder & Harmsen, 2016). It is important to note that more recently, secondary investigations of PISA scores from 2015 have indicated a negative association between students' scientific literacy scores on PISA and the amount of scientific inquiry teaching that students report in their classrooms, on the PISA questionnaire (Cairns & Areepattamannil, 2019; Oliver et al., 2021). These investigations have led to recommendations that scientific inquiry teaching should be curtailed in classrooms (Cairns & Areepattamannil, 2022). However, we would agree with Sjøberg (2018) who stresses that we should not use higher scores on standardized achievement tests to make decisions on whether or not to include science inquiry teaching in the science curriculum. Sjøberg (2018) makes the very important point that we should be more concerned about the beneficial effect of scientific inquiry teaching on students' developing positive attitudes, critical thinking, engagement, interest and motivation, noting that "a written (or digital) test like PISA can hardly measure the skills and competencies acquired in experimental work in a lab or on an excursion; neither can it capture the kind of interest, curiosity and enthusiasm that may be the result of argumentation, inquiry, and the search for solutions to questions that the students have formulated themselves" (p. 200).

Scientific inquiry teaching is embedded in the British Columbia Science education curriculum where this study took place. It takes the form of curricular competencies which form the cornerstone of the science curriculum (BC Science Curriculum, 2015). However, scientific inquiry has been shown to be difficult for teachers to implement (Capps et al., 2016; Crawford, 2007; Fazio et al., 2010; Steele et al., 2013), especially for elementary teachers who often have little science background (e.g., Kim & Tan, 2011; Yoon et al., 2012).

2.2.2 The Steps to Inquiry Framework (SIF)

The SIF (Pardo & Parker, 2010) was created by a team of teachers in Ontario; it was based on Buttemer's (2006) inquiry boards and ideas about how to support student-centered science investigations (Bell et al., 2005; Goldworthy & Feasey, 1997) and gradual release of responsibility during inquiry (Bell et al., 2005; Whitworth et al., 2013). SIF is intended to support teachers and students with enacting science inquiry using step-by-step posters and student pages. It guides teachers to listen to and record their students' ideas, thereby moving teachers toward student-centered instruction. The beginning level posters that were used by the teachers in our co-teaching study are shown in Fig. 2.1. These posters are freely available in English and French (Youth Science Canada, n.d.). We had previously studied SIF implementation after 2-day workshops and found that, despite initial enthusiasm, few teachers actually implemented the SIF (Alexander et al., 2018; Rees & Roth, 2017; Rees et al., 2013). We decided to study how a co-teaching professional development experience might better support teachers in implementing SIF and developing dialogic discourse.

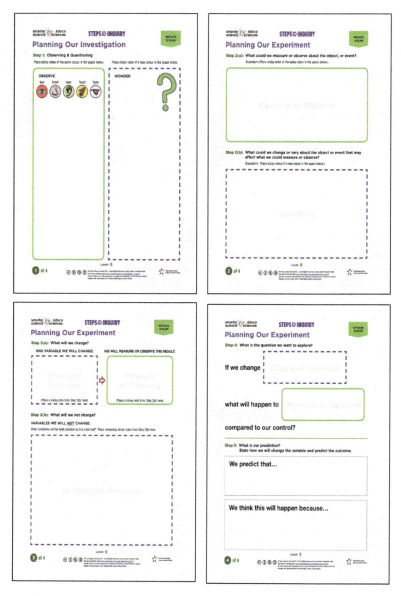

Fig. 2.1 Four SIF posters for planning investigations and experiments (Youth Science Canada, n.d.)

2.2.3 Co-teaching

Our research was centered around a co-teaching model in which two teachers worked in a classroom and shared responsibility for student learning, co-planning, co-teaching, and co-reflecting on student learning (Murphy, 2016; Roth & Tobin, 2002). Although co-teaching is most known for its use in special education (Friend et al., 1993; Harbort et al., 2007; Pancsofar & Petroff, 2016), its use in initial science teacher education has been well documented (e.g., Murphy, 2016; Rees et al., 2022; Roth & Tobin, 2002; Scantlebury et al., 2008); it has also been used for in-service science teacher education (e.g., Roth et al., 1999).

The co-teaching approach used in this study involved a gradual release of responsibility from the expert teacher to the novice teacher through three phases: I Do, We Do, You Do (Duke & Pearson, 2002). In October, the expert teacher observed the novice teacher's usual practice. In November, the expert teacher demonstrated a full SIF-supported science unit with the novice teacher assisting—the I Do phase. In January, the novice teacher conducted a full SIF-supported science unit with the expert teacher assisting—the We Do phase. In February, the novice teacher conducted a full SIF-supported science unit on his own—the You Do phase.

2.2.4 Discourse Patterns

We were interested in exploring how discourse patterns changed during the co-teaching process. The interactions between teachers and students in classrooms has been studied since the 1970s, and it is evident that particular discourse patterns are associated with teacher-directed and in student-centered interactions. Two of the most common teacher-directed discourse patterns in science classrooms are the Initiation-Response-Evaluation (I-R-E) and choral response (Lemke, 1990; Mehan, 1979) while student-centered interactions tend to be more dialogic (Scott et al., 2006).

In the I-R-E pattern, the teacher initiates the interaction with a question, a student responds, and the teacher evaluates, as shown in Turns 1–3 in Table 2.1. Sociolinguists agree that overuse of the I-R-E pattern in classrooms can present a barrier for student learning (Cazden, 2001; Mercer & Dawes, 2014). It can limit students to speaking only when answering test-like questions that teachers provide and evaluate, it can result in a situation where teachers talk on average two-thirds of the time, it can prevent students from deciding when to speak, and it can inhibit students from speaking directly to each other.

A related teacher-directed discourse is choral response, where the whole class responds as a group to a prompt from the teacher, as shown in Turns 4 and 5 in Table 2.1 (Pontefract & Hardman, 2005). Most often used for recall of knowledge, choral response has been seen as suitable for reinforcing knowledge such as decoding, wordlists, and number facts (Rosenshine, 1983). As with I-R-E, overuse of this discourse form can present barriers to students' learning.

Table 2.1 Example of the Phases in I-R-E and Choral response discourse patterns

Turn	Phase	Speaker	
1	Initiation	Mr. Holmes	What is Stage 2, Emma?
2	Response	Emma	Sprout
3	Evaluation	Mr. Holmes	Sprout
4	Prompt	Mr. Holmes	Everybody?
5	Chorus	Students (in unison)	Sprout!

Table 2.2 Example of a student-centered dialogic discourse pattern

Phase	Speaker	
Initiation	Teacher	So, what did you change?
Response	Student	I changed metal, I put metal on there so then I would see how fast it goes
Feedback	Teacher	Yeah
Response	Student	And Brian had two cars
Feedback	Teacher	Yeah
Response	Student	And I and then and, and one of them goes faster and Brian …
Initiation	Teacher	You talk about yours. You changed what it rolled on the bottom?
Response	Student	Yeah
Initiation	Teacher	And what happened to how fast it went?
Response	Student	Um, it goes faster
Initiation	Teacher	On the metal or on the normal floor?
Response	Student	On the metal
Evaluation	Teacher	Okay (nodding), that's a very interesting result, thank you

In contrast to teacher-directed discourse patterns, student-centred discourse pattern involves students responding to open-ended questions (Alexander, 2010; Christoph & Nystrand, 2001). This pattern is sometimes called a dialogic pattern of discourse (Scott et al., 2006) that can be represented as I-R-F-R-F, where F indicates feedback (Mortimer & Scott, 2003). Chains of dialogue will flow and are cumulative (Alexander, 2010); responses are followed and built upon, as shown in Table 2.2.

One component of SIF-supported scientific inquiry is an emphasis on student-centered discourse. In our research, we examined how a novice teacher acquired this discourse pattern through co-teaching, a model that provided multiple opportunities to ask questions and receive just-in-time suggestions. Thus, the novice teacher had the opportunity to alter his practice in the moment to adapt and improve his approach to science inquiry.

2.3   Methods

This qualitative case study (Yin, 2018) looked at both the science education activities and the teacher–student discourse that took place in a Grade 1 classroom: first, before the co-teaching experience and then through the three phases of co-teaching. This study also examined co-teacher interactions to see how teaching with the expert teacher supported the novice teacher in his classroom.

2.3.1   Context

The setting of this study was an elementary school in a small city in Western Canada. This public K–6 school was a school of choice, meaning that students from anywhere in the city could choose to attend if they wanted to focus on science and technology. The school had an inquiry-based teaching philosophy. The school served a high proportion of low-income families.

2.3.2   Participants

Participants included two teachers and 17 children: Mr. Wise (the expert teacher), Mr. Holmes (the novice teacher), and Mr. Holmes' Grade 1 students (all names are pseudonyms). Mr. Wise held a bachelor's degree in science and a master's degree in education and had worked at the school for 8 years. He had experience conducting SIF-supported scientific inquiry for 4 years in his Grade 6 classroom and aiding other elementary teachers doing SIF-supported scientific inquiry for 2 years. In this chapter, we focus on his support of Mr. Holmes, a teacher new to the school who had little experience with teaching science. Mr. Holmes had a bachelor's degree in language arts and had 4 years' teaching experience. The 17 students were aged 6 to 7 years old. Following approvals from the university research ethics board, the school district, and the school principal, informed consent letters were sent to the students' parents and guardians inviting their participation in the study; all agreed to do so.

2.3.3   Data Collection and Analysis

Data for this study consisted of approximately 12 h of video and audio recordings and approximately 100 photographs. Video and audio recordings were taken in the classroom before SIF-supported scientific inquiry was introduced (3 days for 1 h each day) and through the three phases of co-teaching (3 units of approximately 3 h each).

Following existing recommendations for data collection (Roth & Hsu, 2010), one fixed camera video-recorded the whole class, and two handheld cameras followed the two teachers. Audio recorders were set on student tables to capture dialogue that might be missed in the video recordings. The aim was to record, as much as possible, all activities and discourse in the classroom. In addition, we photographed the SIF posters and students' work in booklets.

Data analysis included creating running records (see example in the Appendix) and conducting interaction analysis (Jordan & Henderson, 1995) of teacher–student discourse. We independently constructed running records of video recordings and discussed any issues of interpretation until consensus was reached. The running records included information about classroom events and subevents; examples are teacher organizing students in large- or small-group activities or students conducting scientific inquiry activities, such as collecting observations, developing wonderings, identifying variables, and completing experiments. Photographs of SIF posters and students' work in booklets were used to augment the descriptions of events and deepen our understanding of classroom activities.

To conduct the interaction analysis of teacher–student discourse, we first transcribed the videos taken with the two cameras focused on the teachers, using the fixed-camera videos and audio recordings to fill in gaps and create a verbatim transcript. Next, we worked as a team to examine the videos, read the transcripts, then code the teacher–student discourse patterns as I-R-E, choral response, or dialogic. We used the running records to provide context for the occurrence of discourse patterns.

2.4   Results

To answer our research question *How did discourse patterns change in a Grade 1 science classroom throughout a co-teaching experience?* We begin by describing the classroom activities and discourse patterns that we observed before co-teaching began. We present each of the I Do, We Do, and You Do phases of the co-teaching model in the same way.

2.4.1   October: Before Co-teaching Began

In October, before the co-teaching began and the SIF-supported scientific inquiry was introduced, Mr. Holmes chose the topics of pumpkins, life cycles, and dinosaurs based on the science curriculum, the time of year, the students' interests, and available materials. During our visits, the desks were in rows on one side of the room and the students remained in their desk most of the time. Mr. Holmes mainly taught from the front of the classroom, either standing or sitting on a stool, and circulating at times to give out and retrieve paper. The students were quiet and demonstrated care in following Mr. Holmes' instructions. For example, when preparing to do some deskwork, students needed to retrieve their pencils from their cubbies. Mr. Holmes called, in turn, to the leader of each of the four rows, to lead their row quietly to the cubbies and return to their seats.

Near the beginning of the pumpkin science unit, Mr. Holmes asked the students to recall what they had previously observed about pumpkins. They raised their hands, and he selected a couple of them to respond. Next, he read a book about pumpkins, talked about pumpkin patches, and asked a variety of closed and open-ended questions. Then he showed a video about the life cycle of pumpkin plants, asking questions throughout, and followed up by sharing a story about pumpkins and composting. He showed a pumpkin and invited them to think-pair-share about what they still wondered about pumpkins. He wrote some of their wonderings on the smartboard then asked them to write one of their wonderings on a sticky note to put in their journal. Finally, he asked them to divide a page into quadrants then draw and label the four stages of the pumpkin plant life cycle. He wrote the words *flower*, *seed*, *sprout*, and *pumpkin* on the smartboard for them to copy.

2.4.2   Discourse Patterns Before Co-teaching

During our October visits, the two most prominent discourse patterns that we observed were the choral response (Pontefract & Hardman, 2005) and the I-R-E (Mehan, 1979), examples of which are shown in Table 2.1. In these patterns, Mr. Holmes did most of the talking and students answered questions with one or two words. We did see some evidence of a discourse pattern similar to dialogic (Scott et al., 2006) when Mr. Holmes asked, "What do you remember about pumpkins?" The students' responses were longer and the turns were more cumulative than in a standard I-R-E pattern. In total, during the three days of video recordings, the frequencies of the three discourse patterns we observed were choral response (42%), I-R-E (37%), and dialogic (20%).

2.4.3   November: I Do Co-teaching Phase

Mr. Wise and Mr. Holmes had previously met to plan their first SIF-supported unit on marbles and ramps. With Mr. Holmes assisting, Mr. Wise would lead three 1-hr classes using equipment that included marbles, ramps, and blocks of different sizes and materials. When we arrived for the first class in November, the students were sitting in a circle on the floor at the front of the room. Mr. Wise was standing beside three SIF posters that were on the wall, and Mr. Holmes was sitting beside him. Mr. Wise introduced the unit by talking about making observations using their senses. He showed the materials that would be used and demonstrated rolling a marble down a ramp. He showed the booklets with the student pages that would be used to record their observations. Mr. Holmes organized the students into pairs and helped Mr. Wise distribute the basic materials. The students had 10 min to set up a ramp and roll a marble down it, making observations and recording those observations with drawings and words in their booklet.

Mr. Wise then asked the students to bring their booklet and sit in a circle in front of the room. Once they were settled, Mr. Wise led the class in sharing observations that he and Mr. Holmes wrote on sticky notes and attached to the poster in the section labeled observations (Fig. 2.2). Then Mr. Wise talked about wondering; he explained that after scientists make observations they take time to wonder about those observations. The students returned to where they had been working and recorded their wonderings in their booklet. At the end of the class, the teachers collected the booklets and the students returned their materials.

When we returned for the second lesson, the students were sitting in their desks and Mr. Wise was at the front of the room. He began by reminding them about observing and wondering, then he and Mr. Holmes distributed booklets and asked

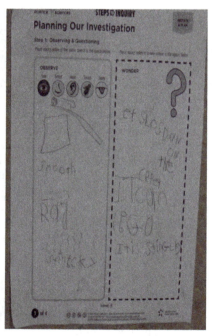

Fig. 2.2 An Example of a Completed SIF Poster (left) and a Related Student Page (right) *Note* The coloured sticky notes on the SIF poster (left) are the students' responses to the question *What did you observe?* Each colour represents observations made using a different sense (e.g., sight, smell). The yellow sticky notes are their responses to the question *What did you wonder?* At the top of the student page (right) in the observe section is a student's diagram of the block, ramp, and marble. Observations include the words "smooth ra[m]p" and "stincky." In the wonder section are the words "et slosdun on the crpet" and "itcan go it is stingcky" that we interpreted as "I wonder if it [the marble] slows down on the carpet" and "I wonder if it [the marble] can go if it is sticky"

them to sit with their partner to talk about their wonders from the last class. A little later when Mr. Wise asked them to come sit in a circle at the front of the room, he asked them to share their wonders, which he and Mr. Holmes wrote on sticky notes and attached to the poster. Mr. Wise told the students that they would now try out some of their wonders. He and Mr. Holmes handed out equipment. Students explored for 8 min before Mr. Wise asked them to leave their equipment and return to the circle. Mr. Wise asked the students to tell what they had changed and what had happened. He and Mr. Holmes recorded responses, which Mr. Wise explained were variables, on sticky notes that were placed on the second SIF poster. Then they returned to their spot to write or draw things that could be variables. At the end of the class, they returned their materials and Mr. Holmes collected their booklets.

At the beginning of the final lesson of the unit, the students were sitting in a circle at the front of the room. Mr. Wise stood by the posters and showed how he would plan an experiment by moving sticky notes from Poster 2 to Poster 3. He indicated one thing that he might change and one thing that he might measure as well as things that he would need to keep the same. Mr. Wise and Mr. Holmes then gave them their booklets and asked them to find a spot on the floor to work while planning their experiment—what they would change, what they would measure, and what they would keep the same. When they had a plan, they went to Mr. Wise, told him the plan, and were given the materials they needed. They had 15 min to conduct their experiment before Mr. Wise asked them to return their materials and sit in a circle at the front of the class. Mr. Wise asked each pair what they had changed and what they had measured.

2.4.4 Discourse Patterns in the I Do Co-teaching Phase

During our visits in the I Do phase, we observed I-R-E and choral response patterns when Mr. Wise introduced activities to the students at the beginning of each class. We observed the dialogic pattern when he asked students to share their observations, their wonderings, and their ideas of variables to change and measure, and when he asked them about their experiments. The relative frequency of the patterns was different than in the unit prior to co-teaching: dialogic was the most frequent (71%), I-R-E was next (28%), and choral response was rarely observed (1%).

2.4.5 January: We Do Co-teaching Phase

Prior to our visit in January for the We Do co-teaching phase, Mr. Wise and Mr. Holmes had met to plan the second SIF-supported unit that involved cars and tracks. Mr. Wise suggested the unit because he had done it before. He had already gathered materials so that groups would be able to choose sizes and shapes of cars; lengths, widths, and materials of track; and materials for the car to land on. Mr. Holmes would lead the unit and Mr. Wise would assist when needed.

When we arrived, the students were sitting in a circle at the front of the room. Mr. Holmes was standing beside the three SIF posters, and Mr. Wise was standing at the side of the room. Mr. Holmes asked Mr. Wise what to do first; Mr. Wise suggested starting with a review about making observations, then distributing booklets and materials so that the students could make observations. However, Mr. Holmes jumped ahead in that plan and asked what they wondered about and what variables they could change. Mr. Wise spoke up and suggested to Mr. Holmes that they needed to start with observations. Because some students had their hands up already to answer Mr. Holmes' question about wondering, he took one question. Then, following Mr. Wise's advice, he explained that first they would be observing what happened with the ramp and car. They were given 10 min to work in pairs at spots around the room to set up their track, roll the cars, and record their observations. Both teachers circulated to provide support as needed. The teachers called the students to bring their booklets back to the circle, where Mr. Holmes sat on a stool and Mr. Wise stood beside him. Mr. Holmes began by asking what they had found out; however, Mr. Wise stepped in to ask instead what they had observed, which had been the task.

Following Mr. Wise's correction, Mr. Holmes asked what they had observed while Mr. Wise wrote the observations on sticky notes and attached them to the poster. Then Mr. Holmes, with assistance from Mr. Wise, explained that they would go back to their spots to talk about and record their wonderings. Both teachers circulated to assist as needed. For the rest of the class and the remaining two classes of the unit, the teachers worked together to follow the SIF as Mr. Wise had during the I Do phase. Mr. Holmes took the lead but asked Mr. Wise for help about what to do next and adjusted his plans accordingly.

2.4.6 Discourse Patterns in the We Do Co-teaching Phase

In the We Do phase, the three discourse patterns were again evident. We observed the I-R-E pattern when Mr. Holmes introduced activities and the choral response pattern when he asked students to repeat what was written on the SIF posters. We noted that Mr. Wise and Mr. Holmes worked together to generate dialogic discourse. The frequencies of discourse patterns were similar to the frequencies observed in the I Do phase: dialogic (80%), I-R-E (15%), choral response (5%).

In this phase of the co-teaching model, we were interested in the dialogic interactions between the teachers. Our analysis showed that Mr. Holmes and Mr. Wise interacted frequently throughout the We Do phase of co-teaching. During the 117 min of video-recorded class-time, we noted 88 interactions that we categorized as shown in Table 2.3.

Table 2.3 Types of dialogic interactions between two teachers in the We Do phase

Type of interaction	Description	Example	Frequency
Aside	Teachers speak privately to each other, quietly and quickly	Mr. Wise said quickly and quietly to Mr. Holmes, "Probably you should demo this" indicating a step on the SIF poster that dealt with variables	27
Check-in	One teacher checks in with the other teacher (e.g., regarding what is coming up)	Mr. Holmes at one point said to Mr. Wise, "What do you think, maybe one more minute?"	25
Performance for students	Teachers speak to one another more loudly, more deliberately, and at a slower pace	Mr. Wise said slowly and in a loud voice, "So, what did we learn, Mr. Holmes?" Mr. Holmes responded, "I learned that if it [the track] gets too steep … it causes it [the car] to tumble."	15
Interjection	One teacher speaks up when the other teacher is leading	Mr. Wise interrupted Mr. Holmes, indicating that some clarification was needed	14
		Total	81

2.4.7   *February: You Do Co-teaching Phase*

Prior to our visit in February for the You Do phase, Mr. Wise and Mr. Holmes met to plan the third SIF-supported unit. Mr. Holmes suggested the topic of magnets and together they chose a magnetic kite activity that involved a paper clip attached to a string and a magnet used to make the paper clip travel through the air without touching it. Mr. Holmes had gathered the materials so that each group would be able to choose from a range of sizes of paper clips, string of various thickness and length, and magnets of different sizes and strengths. The teachers decided that although Mr. Wise would not join the class students from his Grade 6 class would join to assist the Grade 1 students.

When we arrived for the You Do phase, the students were sitting in a circle and Mr. Holmes was sitting beside the three SIF posters. He explained that they would be doing a new experiment and reminded them about making observations using their senses. He then explained the magnet activity and introduced the Grade 6 students who had joined the class to help. Mr. Holmes handed out booklets and materials; the Grade 1 students had 15 min to make their observations while he and the Grade 6 students circulated to help. Then he called the students back to the circle to share their observations. Over the next 2 days of the unit, Mr. Holmes continued to follow the SIF as Mr. Wise had during the I Do phase and both had during the We Do phase.

2.4.8 Discourse Patterns in the You Do Co-teaching Phase

All three discourse patterns were evident during the You Do phase. The dialogic pattern occurred during the times when the students were in a circle sharing what they had observed, wondered, and found out during their experiments. As in the I Do and We Do phases, we noticed the I-R-E pattern when Mr. Holmes introduced activities. The choral response pattern was relatively rare but was used when, for example, Mr. Holmes prompted the children to tell him the five senses. The frequencies of discourse patterns we observed in the I Do phase were dialogic (71%), I-R-E (26%), and choral response (2%).

2.4.9 Discussion

We found a substantial shift in discourse patterns before and during co-teaching. Through the I Do and We Do phases of co-teaching, Mr. Wise worked together with Mr. Holmes to implement SIF-supported units and to encourage student-centered dialogue (e.g., dialogic discourse patterns). In the You Do phase, Mr. Holmes implemented a SIF-supported activity on his own and was able to foster student-centered dialogue (see Table 2.4).

Examining the dialogic interactions between the co-teachers in the We Do phase, we found an average of one interaction every 1 to 2 min. Mr. Holmes was learning in the moment to implement SIF-supported science inquiry and associated dialogic discourse patterns with his students. During these interactions, Mr. Wise helped keep Mr. Holmes on track by offering corrections and suggestions; and Mr. Holmes frequently asked Mr. Wise questions. The co-teachers' interactions in our study fit the description of *huddles*, defined by Soslau et al. (2018) as short, focused meetings where teachers can learn from each other before, during, or after a lesson. Huddles can be used to help teachers engage in a particular strategy, develop their questioning and pacing, manage the classroom as well as model, provide corrections and enhancements, and clarify directions. In this study, the dialogic interactions could

Table 2.4 Discourse patterns before and during co-teaching

Phase		Who was teaching?	Discourse pattern		
			I-R-E (%)	Choral response (%)	Dialogic (%)
Before co-teaching		Novice teacher, on his own	37	42	20
Co-teaching	I Do	Mentor teacher, on his own	28	1	71
	We Do	Co-teaching	15	5	80
	You Do	Novice teacher, on his own	26	2	71

be viewed as huddles where the expert teacher helped the novice teacher implement SIF-supported teaching and discourse.

2.5   Conclusions

This study was designed to address the research question: How did discourse patterns change in a Grade 1 science classroom throughout a co-teaching experience? In the co-teaching model that we used, Mr. Holmes had first observed and assisted Mr. Wise in the I Do phase as Mr. Wise modelled SIF-supported scientific inquiry and associated dialogic discourse patterns. During the We Do phase, Mr. Holmes led a unit of SIF-supported scientific inquiry with the assistance of Mr. Wise. The two teachers interacted frequently (e.g., brief interactions or huddles); Mr. Holmes was able to learn in the moment by asking questions of Mr. Wise, who offered suggestions when they were most helpful. During the You Do phase, Mr. Holmes was able to build on his experiences during the I Do and We Do phases, implementing SIF-supported science inquiry and student-centered dialogic discourse with his students.

We found that by the end of the co-teaching professional learning experience Mr. Holmes was able to move toward the dialogic patterns associated with student-centered discourse. He led the class through the steps of observation then developing wonderings, planning and conducting an experiment, and communicating findings. His SIF-supported unit included frequent opportunities for dialogic interactions where students had opportunities to talk and share their thinking about their activities. The dialogic pattern was the most common pattern during the You Do phase, just as it had been in the I Do phase modelled by Mr. Wise and in the We Do phase when both teachers worked together. Although this qualitative case study was limited to one classroom, our findings support the idea that the I Do, We Do, You Do co-teaching model can be an effective approach to teacher professional learning.

Acknowledgements Thank you to Robert Wielgoz and Justin Hummel for their extensive work and support. Thank you to Richard Pardo, Jennifer Parker, Michael Newnham, and other teachers for their work on the creation of the Steps to Inquiry Framework. This research was supported by the Social Sciences and Humanities Research Council of Canada.

Appendix

Running Record: Before Co-teaching, Class 2, total time: 38.30 minutes

Interval (min)	Activity: What is Mr. Holmes doing? What are the students doing?
00:00–1:40	Mr. Holmes, standing in front of class, asks students what they remember from the last class. Students are sitting in their seats facing front, raising hands to respond. He either calls on students to answer by name or by pointing to a student
1:40–04:05	Mr. Holmes, standing in front of class, introduces a book entitled *My Pumpkin*. While reading the book, he asks known-answer questions and/or provides prompts. Students are sitting in their seats facing front. They respond as required, either raising hand to answer or responding to prompt in chorus
04:05–6:46	Mr. Holmes, standing in front of class, talks about pumpkin patches. He asks a mix of known-answer and open questions about pumpkin patches and what it means to be a living thing. Students are sitting in their seats facing front. They respond as required, either raising hand to answer or responding to prompt in chorus
06:46 – 07:15	Mr. Holmes moves in the classroom to turn off the light and set up the video. Students are sitting in their seats facing front
07:15–12:20	Mr. Holmes is kneeling as the video is playing. He makes comments and asks students questions or provides prompts about the video. Students are sitting in their seats facing front. They raise hands and respond to known-answer questions as required and respond to prompts in chorus. Sometimes he pauses the video for the questions and prompts
12:20–16:35	Mr. Holmes stands up in front of the class, then moves to the side to carry a pumpkin, then back to the front again, sometimes moves to the middle too. He talks about topics related to pumpkins, pumpkin patches, and shares his story about composting. He asks known-answer questions to which the students respond. Students are sitting in their seats facing front or turning to the side. Lights are still off. Video is finished
16:35–17:37	Mr. Holmes turns the lights on and moves between the side of the classroom and the front then to the back. He asks students to turn to a partner and speak about one thing they are still wondering about in relation to pumpkins (Think-Pair-Share). Students are sitting in their seats, turn and speak with a partner
17:37–23:23	Mr. Holmes moves to the front then goes to the smart board. Students are sitting in their seats facing front. He asks them to share what they are wondering about pumpkins. They raise their hands. He indicates to students to answer, sometimes using their name or by pointing to a student. Students share their wonderings or what their questions are. He repeats their comments and sometimes adds some remarks. He writes some of the wonderings on the smartboard
23:23–28:30	Mr. Holmes is standing in front of the class. Students are sitting in their seats facing front. A designated helper student takes sticky notes and passes them to all the students. They already have pencils. He asks them to write their "wonder" questions on the sticky notes. He answers some questions about spelling and makes a few comments to remind some students of what they wrote. He occasionally points to words he has written on the smartboard so that they can copy them
28:30–33:50	Mr. Holmes walks around the classroom instructing students who are finished to go to their cubby, bring back their journal, and put their sticky notes inside it. Students are sitting in their seats facing front then start moving to get journals and go back to their seats

(continued)

(continued)

Interval (min)	Activity: What is Mr. Holmes doing? What are the students doing?
33:50–38:10	Mr. Holmes shows students how to divide their page in four with a line in the middle and then to put another line across the middle creating four boxes for the stages of the pumpkin: flower, seed, sprout, and pumpkin. Students sit in their seats and do the task, sometimes asking questions about instructions and spelling. He repeats instructions and assists students. He puts the book on top of the smartboard to show an example
38.10–38.30	There is a fire drill. The bell rings. Mr. Holmes tells the students to stand up and make a line. Students stand next to their seats. (The class exits the building to the playground for the fire drill.)

References

American Association for the Advancement of Science. (1993/2009). *The nature of science.* http://www.project2061.org/publications/bsl/online/index.php?chapter=1

Aditomo, A., & Klieme, E. (2020). Forms of inquiry-based science instruction and their relations with learning outcomes: Evidence from high and low-performing education systems. *International Journal of Science Education, 42*(4), 504–525. https://doi.org/10.1080/09500693.2020.1716093

Alexander, R. (2010). *Dialogic teaching essentials.* https://www.nie.edu.sg/docs/default-source/event-document/final-dialogic-teaching-essentials.pdf

Alexander, E., Pardo, R. Lindsay, S., & Rees, C. A. (2018). The DRiVe inquiry framework. *Alberta Science Education Journal, 45*(3), 29–34. https://sc.teachers.ab.ca/SiteCollectionDocuments/ASEJ/ASEJ-Vol45No3.pdf

Anderson, R. D. (2002). Reforming science teaching: What research says about inquiry. *Journal of Science Teacher Education, 13*(1), 1–12. https://doi.org/10.1023/A:1015171124982

Bell, R. L., Smetana, L., & Binns, I. (2005). Simplifying inquiry instruction. *The Science Teacher, 72*(7), 30–33. http://www.jstor.org/stable/24138115

British Columbia Ministry of Education. (2015). Curricular competencies. https://curriculum.gov.bc.ca/sites/curriculum.gov.bc.ca/files/curriculum/continuous-views/en_science_k-10_curricular-competencies.pdf

British Columbia Ministry of Education. (2016). *British Columbia science curriculum K-12.* https://curriculum.gov.bc.ca/curriculum/science

Buttemer, H. (2006). Inquiry on board! *Science & Children, 44*(2), 34–9. http://www.jstor.org/stable/43172847

Cairns, D., & Areepattamannil, S. (2019). Exploring the relations of inquiry-based teaching to science achievement and dispositions in 54 countries. *Research in Science Education, 49*(1), 1–23.

Cairns, D., & Areepattamannil, S. (2022). Teacher-directed learning approaches and science achievement: Investigating the importance of instructional explanations in Australian schools. *Research in Science Education, 52*(4), 1171–1185.

Capps, D. K., Shemwell, J. T., & Young, A. M. (2016). Over reported and misunderstood? A study of teachers' reported enactment and knowledge of inquiry-based science teaching. *International Journal of Science Education, 38*(6), 934–959. https://doi.org/10.1080/09500693.2016.1173261

Cazden, C. B. (2001). *Classroom discourse: The language of teaching and learning* (2nd ed.). Pearson Education Canada.

Christoph, J. N., & Nystrand, M. (2001). Taking risks, negotiating relationships: "One teacher's transition toward a dialogic classroom." *Research in the Teaching of English, 36*(2), 249–286. http://www.jstor.org/stable/40171538

Crawford, B. A. (2007). Learning to teach science as inquiry in the rough and tumble of practice. *Journal of Research in Science Teaching, 44*(4), 613–642. https://doi.org/10.1002/tea.20157

Dewey, J. (1910). Science as subject-matter and as method. *Science, 31*(787), 121–127. https://doi.org/10.1126/science.31.787.121

Dewey, J. (1938). *Experience and education.* Macmillan.

Duke, N. K., & Pearson, P. D. (2002). Effective practices for developing reading comprehension. In A. E. Farstrup & S. J. Samuels (Eds.), *What research has to say about reading instruction* (3rd ed., pp. 205–242). International Reading Association.

Fazio, X., Melville, W., & Bartley, A. (2010). The problematic nature of the practicum: A key determinant of pre-service teachers' emerging inquiry-based science practices. *Journal of Science Teacher Education, 21*(6), 665–681. https://doi.org/10.1007/s10972-010-9209-9

Friend, M., Reising, M., & Cook, L. (1993). Co-teaching: An overview of the past, a glimpse at the present, and considerations for the future. *Preventing School Failure: Alternative Education for Children & Youth, 37*(4), 6–10. https://doi.org/10.1080/1045988X.1993.9944611

Furtak, E. M., Seidel, T., Iverson, H., & Briggs, D. C. (2012). Experimental and quasi-experimental studies of inquiry-based science teaching: A meta-analysis. *Review of Educational Research, 82*(3), 300–329. https://doi.org/10.3102/0034654312457206

Goldworthy, A., & Feasey, R. (1997). *Making sense of primary science investigations.* Association for Science Education.

Harbort, G., Gunter, P. L., Hull, K., Brown, Q., Venn, M. L., Wiley, L. P., & Wiley, E. W. (2007). Behaviors of teachers in co-taught classes in a secondary school. *Teacher Education & Special Education, 30*(1), 13–23. https://doi.org/10.1177/088840640703000102

Hodson, D. (1996). Laboratory work as scientific method: Three decades of confusion and distortion. *Journal of Curriculum Studies, 28*(2), 115–135. https://doi.org/10.1080/0022027980280201

Hodson, D. (2014). Learning science, learning about science, doing science: Different goals demand different learning methods. *International Journal of Science Education, 36*(15), 2534–2553. https://doi.org/10.1080/09500693.2014.899722

Jordan, B., & Henderson, A. (1995). Interaction analysis: Foundations and practice. *Journal of the Learning Sciences, 4*(1), 39–103. https://doi.org/10.1207/s15327809jls0401_2

Kang, J., & Keinonen, T. (2018). The effect of student-centered approaches on students' interest and achievement in science: Relevant topic-based, open and guided inquiry-based, and discussion-based approaches. *Research in Science Education, 48*(4), 865–885. https://doi.org/10.1007/s11165-016-9590-2

Kim, M., & Tan, A. L. (2011). Rethinking difficulties of teaching inquiry-based practical work: Stories from elementary pre-service teachers. *International Journal of Science Education, 33*(4), 465–486. https://doi.org/10.1080/09500693.2014.899722

Lazonder, A. W., & Harmsen, R. (2016). Meta-analysis of inquiry-based learning: Effects of guidance. *Review of Educational Research, 86*(3), 681–718. https://doi.org/10.3102/0034654315627366

Lehesvuori, S., Viiri, J., & Rasku-Puttonen, H. (2011). Introducing dialogic teaching to science student teachers. *Journal of Science Teacher Education, 22*(8), 705–727. https://doi.org/10.1007/s10972-011-9253-0

Lemke, J. L. (1990). *Talking science: Language, learning, and values.* Ablex.

Mehan, H. (1979). "What time is it, Denise?": Asking known information questions in classroom discourse. *Theory into Practice, 18*(4), 285–294. https://doi.org/10.1080/00405847909542846

Mercer, N., & Dawes, L. (2014). The study of talk between teachers and students, from the 1970s until the 2010s. *Oxford Review of Education, 40*(4), 430–445. https://doi.org/10.1080/03054985.2014.934087

Minner, D. D., Levy, A. J., & Century, J. (2010). Inquiry-based science instruction—what is it and does it matter? Results from a research synthesis years 1984 to 2002. *Journal of Research in Science Teaching, 47*(4), 474–496. https://doi.org/10.1002/tea.20347

Mortimer, E. F., & Scott, P. H. (2003). *Meaning making in secondary science classrooms.* Open University Press.

Murphy, C. (2016). *Coteaching in teacher education: Innovative pedagogy for excellence.* Critical.

National Science Teachers Association. (1971). NSTA position statement on school science education for the 70s. *The Science Teacher, 38*(8), 46–51. https://www.jstor.org/stable/24132493

Oliver, M., McConney, A., & Woods-McConney, A. (2021). The efficacy of inquiry-based instruction in science: A comparative analysis of six countries using PISA 2015. *Research in Science Education, 51*(2), 595–616.

Ontario Ministry of Education. (2007). *The Ontario curriculum grades 1–8: Science and technology.* http://www.edu.gov.on.ca/eng/curriculum/elementary/scientec18currb.pdf

Pancsofar, N., & Petroff, J. G. (2016). Teachers' experiences with co-teaching as a model for inclusive education. *International Journal of Inclusive Education, 20*(10), 1043–1053. https://doi.org/10.1080/13603116.2016.1145264

Pardo, R., & Parker, J. (2010). The inquiry flame: Scaffolding for scientific inquiry through experimental design. *The Science Teacher, 77*(8), 44–49. http://www.jstor.org/stable/24146181

Pontefract, C., & Hardman, F. (2005). The discourse of classroom interaction in Kenyan primary schools. *Comparative Education, 41*(1), 87–106. https://doi.org/10.1080/03050060500073264

Rees, C., Murphy, C., Kaur, R. D., & Brown, A. (2022). Inquiring together: A pre-participation phase for a coteaching clinical practice model in teacher education. *Teaching and Teacher Education, 117*, 103814.

Rees, C. A., Pardo, R., & Parker, J. (2013). Steps to opening science inquiry: Pre-service teachers' practicum experiences with a new support framework. *Journal of Science Teacher Education, 24*(3), 475–496. https://doi.org/10.1007/s10972-012-9315-y

Rees, C. A., & Roth, W.-M. (2017). Interchangeable positions in interaction sequences in science classrooms. *Dialogical Pedagogy, 5.* http://dx.doi.org/10.5195/dpj.2017.184

Rocard, M., Csermely, P., Jorde, D., Lenzen, D., Walberg-Henriksson, H., & Hemmo, V. (2007). *Science education now: A renewed pedagogy for the future of Europe.* https://www.eesc.europa.eu/sites/default/files/resources/docs/rapportrocardfinal.pdf

Rosenshine, B. (1983). Teaching functions in instructional programs. *The Elementary School Journal, 83*(3), 335–351. https://doi.org/10.1086/461321

Roth, W.-M., & Bowen, G. M. (1995). Knowing and interacting: A study of culture, practices, and resources in a grade 8 open-inquiry science classroom guided by a cognitive apprenticeship metaphor. *Cognition & Instruction, 13*(1), 73–128. https://doi.org/10.1207/s1532690xci1301_3

Roth, W. M., & Hsu, P. L. (2010). *Analyzing communication: Praxis of method.* Brill Sense.

Roth, W.-M., Masciotra, D., & Boyd, N. (1999). Becoming-in-the-classroom: A case study of teacher development through coteaching. *Teaching & Teacher Education, 15*(7), 771–784. https://doi.org/10.1016/S0742-051X(99)00027-X

Roth, W.-M., & Tobin, K. G. (2002). *At the elbow of another: Learning to teach by co-teaching.* Peter Lang.

Scantlebury, K., Gallo-Fox, J., & Wassell, B. (2008). Coteaching as a model for preservice secondary science teacher education. *Teaching & Teacher Education, 24*(4), 967–981. https://doi.org/10.1016/j.tate.2007.10.008

Schwab, J. J. (1960). Inquiry, the science teacher, and the educator. *The School Review, 68*(2), 176–195. http://www.jstor.org/stable/1083585

Schwab, J. J. (1962). *The teaching of science as enquiry.* Harvard University Press.

Scott, P. H., Mortimer, E. F., & Aguiar, O. G. (2006). The tension between authoritative and dialogic discourse: A fundamental characteristic of meaning making interactions in high school science lessons. *Science Education, 90*(4), 605–631. https://doi.org/10.1002/sce.20131

Sjøberg, S. (2018). The power and paradoxes of PISA: Should inquiry-based science education be sacrificed to climb on the rankings? *Nordic Studies in Science Education, 14*(2), 186–202.

Soslau, E., Kotch-Jester, S., Scantlebury, K., & Gleason, S. (2018) Coteachers' huddles: Developing adaptive teaching expertise during student teaching. *Teaching & Teacher Education, 73*(1), 99–108. https://doi.org/10.1016/j.tate.2018.03.016

Steele, A., Brew, C., Rees, C. A., & Ibrahim-Khan, S. (2013). Our practice, their readiness: Teacher educators collaborate to explore and improve pre-service teacher readiness for science and math instruction. *Journal of Science Teacher Education, 24*(1), 113–131. https://doi.org/10.1007/s10 972-012-9311-2

Tuan, H.-L., Chin, C.-C., Tsai, C. C., & Cheng, S.-F. (2005). Investigating the effectiveness of inquiry instruction on the motivation of students with different styles. *International Journal of Science & Mathematics Education, 3*(4), 541–566. https://doi.org/10.1007/s10763-004-6827-8

Tytler, R. (2007). *Re-imagining science education: Engaging students in science for Australia's future.* ACER Press.

United States National Research Council. (2013). *Next generation science standards: For states, by states.* National Academies Press. https://doi.org/10.17226/18290

Whitworth, B. A., Maeng, J. L., & Bell, R. L. (2013). Differentiating inquiry. *Science Scope, 37*(2), 10–17. http://www.jstor.org/stable/43827051

Yin, R. K. (2018). *Case study research and applications.* SAGE.

Yoon, H.-G., Joung, Y. J., & Kim, M. (2012). The challenges of science inquiry teaching for pre-service teachers in elementary classrooms: Difficulties on and under the scene. *Research in Science Education, 42*(3), 589–608. https://doi.org/10.1007/s11165-011-9212-y

Youth Science Canada. (n.d.). *Smarter science posters and pdfs.* https://smarterscience.youthscie nce.ca/

Carol A. B. Rees BSc (Trinity College Dublin), Teaching Dip (Montessori St. Nicholas Centre), MA (Mount St. Vincent University), PhD (University of Western Ontario), is a professor of science education in the Faculty of Education & Social Work at Thompson Rivers University. She was a scientist before she became a teacher, which influences her ways of thinking about science education. Her research interests include student-centered scientific inquiry, discourse patterns in classrooms, and co-teaching for teacher professional learning in science education. Current research projects include *Supporting curiosity-driven, inquiry-based science education online through a community of inquiry partnership: rethinking pedagogical approaches during the Covid-19 pandemic; and Supporting student-centered pedagogy and dialogic teaching through co-teaching.*

Chapter 3
Adopting Universal Design for Learning as a Means to Foster Inclusive Science Teaching and Learning

Karen Goodnough, Saiqa Azam, and Patrick Wells

3.1 Introduction

With calls locally, nationally, and internationally for inclusive education reforms (Campbell et al., 2016; Collins et al., 2017; International Disability and Development Consortium, 2017; United Nations Educational, Scientific and Cultural Organization [UNESCO], 2000), examining the role of the teacher in creating inclusive learning environments is critical. Research has shown that teachers' attitudes toward and perspectives about inclusion influence their ability and/or willingness to adopt inclusion in the regular classroom (Aldani, 2020; Leifler, 2020; Materechera, 2020; Saloviita & Schaffus, 2016; Wilson, 2014). A lack of knowledge and understanding about how to foster inclusive learning environments may also present barriers to teachers' adoption of inclusion (Blanton et al., 2011; Krischler et al., 2019; Lambe & Bones, 2006; Materechera, 2020; Stronge et al., 2007). Consequently, having insight into how and why teachers adopt inclusive practices is important if all students are to have access to the curriculum. UDL is a research-based framework that may be used to improve and optimize teaching and learning through the design of learning environments that are effective and accessible to all learners (CAST, 2018; Meyer et al., 2014).

K. Goodnough (✉) · S. Azam · P. Wells
Memorial University of Newfoundland, Faculty of Education, 300 Prince Philip Drive, St. John's, NL A1B 3X8, Canada
e-mail: kareng@mun.ca

S. Azam
e-mail: sazam@mun.ca

P. Wells
e-mail: p.wells@mun.ca

© The Author(s), under exclusive license to Springer Nature Switzerland AG 2023   29
C. D. Tippett and T. M. Milford (eds.), *Exploring Elementary Science Teaching and Learning in Canada*, Contemporary Trends and Issues in Science Education 53, https://doi.org/10.1007/978-3-031-23936-6_3

This case study examines the experiences of a school-based teacher inquiry group consisting of a Grade 2 classroom teacher, an instructional resource teacher, and an assistant principal as they assessed their inclusive practices as framed by UDL. We supported the teachers as they explored how UDL Guidelines and principles could facilitate changes to their classroom practices. The teacher inquiry project described here was guided by two research questions that focused primarily on the professional learning of the classroom teacher:

1. How will the teacher interpret the UDL framework?
2. How will the UDL framework inform the teacher's classroom practice in science?

3.2 Inclusion, UDL, and Science Teaching

The concept of inclusion and the application of its principles in schools and class-rooms is prevalent today as a means to meet the learning needs of all students in the regular classroom. The meaning of inclusion and how it gets enacted can vary, however. Key aspects of inclusion often include finding the best ways for all students to participate fully in school and have access to the curriculum, reducing barriers to full inclusion of all in general education, fostering equity and excellence; promoting a sense of belonging for all in-school communities, and respecting the uniqueness of individuals (Farrell, 2016; Opertti et al., 2009; Shore et al., 2011). The UNESCO (2005) principles of inclusion also align with these ideas:

- Inclusion is a process.
- Inclusion is concerned with the identification and removal of barriers.
- Inclusion is about the presence, participation and achievement of all students.
- Inclusion involves a particular emphasis on those groups of learners who may be at risk of marginalization, exclusion, or underachievement (pp. 15–16).

In terms of teachers' understanding and adoption of inclusive practices, research has shown that teachers struggle with creating inclusive learning environments (Avramidis & Norwich, 2002; Damianidou & Phtiaka, 2018; Florian & Graham, 2014; Sagner-Tapia, 2018). One framework that is being adopted in K–12 settings to assist teachers in creating inclusive learning environments is UDL. Three broad principles underpin the UDL framework:

1. Provide multiple means of **Engagement**
2. Provide multiple means of **Representation**
3. Provide multiple means of **Action and Expression** (CAST, 2018, "UDL Guidelines" section).

Each principle includes three guidelines and several checkpoints that can be used to guide educators in planning curriculum, instruction, and assessment (see http://udl guidelines.cast.org/ for more detail). For example, to offer people multiple means of engagement in learning, the UDL framework suggests that learner choice and autonomy be optimized by offering choices in how to meet a particular learning outcome while considering the nature of the choice and the level of independence required to complete a learning task. Ways to achieve this in the classroom may include helping students set their own learning goals, allowing them to design learning tasks, and/or providing them with choice in tools for completing tasks (CAST, 2020a). Because students learning science may struggle due to how the curriculum is delivered rather than their deficits (Meyer et al., 2014), UDL can be adopted in science to foster inclusivity, particularly through careful attention to how UDL guidelines and principles may inform practice.

While a growing body of research is emerging on how teachers are utilizing UDL to inform classroom praxis, a limited number of studies have been reported in the literature on how UDL has been utilized for teaching and learning in science (e.g., Basham & Marino, 2013; Dymond et al., 2006; Marino, 2010; Riedell, 2016), and especially elementary science. Finnegan and Dieker (2019) described modifying the elementary curriculum by considering principles of multiple means of representation to help students convey their understanding of scientific content. Rappolt-Schlichtmann et al. (2013) designed web-based science notebooks in an elementary context using the UDL framework to examine teacher and student adoption of the notebooks and student learning of content outcomes. In this study, the adoption of the UDL framework was intended to reduce barriers to student learning in science by making the curriculum accessible to all students. This case study examining the creation of inclusive science classrooms through UDL centres on the insights of one teacher who was part of a small inquiry group.

3.3   Methods

In this chapter, we focus on a single case study, which was part of a 5-year Teachers in Action professional development initiative in Newfoundland and Labrador that was funded by a local oil consortium. The main goals of Teachers in Action were to support teachers in becoming more confident in teaching in science, technology, engineering, and mathematics (STEM) subjects and to assist them in adopting inquiry-based approaches in STEM teaching and learning. School-based teams ranging in size from 2 to 8 teachers engaged in action research cycles of collaborative action: planning, acting, observing, and reflecting (Kemmis & McTaggart, 2005). All participating teachers were volunteers; they were provided with 5–7 days of release time per year, a small budget for classroom resources, and mentoring from members of the project leadership team (the first and second authors and a full-time professional development facilitator). This chapter will draw upon data from three participating teachers; however, descriptions of detailed planning and implementation in science will focus on the experiences of one Grade 2 teacher who was part of a school-based inquiry group that had been part of the larger project for three years and was beginning its fourth cycle of action research. Andy, Lisa, and Susan (all names are pseudonyms) worked in a K–5 school of approximately 300 students. The school served families from diverse backgrounds, had strong parental involvement, and was focused on infusing STEM principles into the curriculum.

Andy had 24 years of teaching experience and was an action researcher; she held undergraduate degrees in education and special education and a graduate degree in curriculum, teaching, and learning. She described her Grade 2 class of 22 students during a planning meeting as a "dynamic group … [with] 10 students being serviced by an instructional resource teacher and having identified special needs related to behaviour and academics." She focused her planning and implementation on helping her students become more reflective and aware of their strengths and learning preferences and to become more self-guided science learners. She hoped that by participating in this inquiry group she could make the science curriculum accessible to all students in her class.

Lisa was an instructional resource teacher (IRT) who provided support to classroom teachers by planning with them and supporting identified special needs children inside and outside the regular classroom. She had 10 years teaching experience, mostly as an IRT, and held undergraduate degrees in education and special education and a graduate degree in educational technology. In this project, she chose to focus her inquiry on a kindergarten child with Down's Syndrome who was part of a class of 5- to 6-year-olds with varying abilities.

Susan was an experienced educator of 25 years, who had started a new position as an assistant principal while retaining part-time responsibilities as an IRT teacher. She held undergraduate degrees in education and special education and a graduate degree in leadership.

In the fourth year of Teachers in Action, Andy engaged in a year-long cycle of collaborative action research to understand and adopt UDL principles (e.g., multiple means of engagements, representation, and actions and expressions). In this school-based teacher inquiry, the experiences of these three teachers were examined to interpret their perceptions and views of the UDL framework and its potential as a tool to inform inclusive science education. A case study method (Merriam & Tisdell, 2015) was adopted to allow us to capture the complexity of teacher learning about becoming inclusive science/STEM teachers using UDL. An intense, holistic, descriptive, qualitative exploration was conducted of a "single unit or bounded system" with defined boundaries of one primary/elementary school, three teachers, and a one-year time frame (Merriam, 1998, p. 12). The flexible nature of a case study revealed "holistic characteristics" of natural events and behaviours while the researchers investigated the inclusive practices of Andy (Yin, 1994, p. 23). The teachers worked collaboratively from October to March, meeting regularly for one day per month to plan. Multiple sources of data were collected from October to June, allowing us to develop a robust picture of how the teachers were adopting UDL in a science unit on relative position and motion.

3.3.1   Interviews

We individually interviewed each teacher for one hour at the end of the teacher inquiry project using a semi-structured interview protocol. The intention was to gather data about inclusive practices and the implementation of UDL principles by examining their professional learning in science. Interview questions included: Have your beliefs and values about inclusion and student diversity changed in any way? Can you talk about the UDL principles you used and how they were embedded in your classroom? How do you differentiate between UDL and differentiated instruction (DI)? What are some of the tensions or challenges that still exist? Each interview was audio recorded, each audio file was transcribed verbatim, and the transcripts were used as a data source.

3.3.2   Teacher Reflections

Each teacher wrote a reflection (1,100–2,500 words) after each collaborative planning meeting and teaching day. The teachers were provided with a reflective framework (i.e., What? So what? Now what?) to guide their reflections. In total, 20 reflections were completed during the planning and implementation stages of the teacher inquiry project. The process of reflecting allowed teachers to confront their beliefs about inclusion and document any changes occurring in their beliefs and practices regarding inclusive science education and UDL.

3.3.3   Teacher/Student Artifacts

During the year-long collaborative teacher inquiry project, Andy, Lisa, and Susan created a variety of artifacts (e.g., lesson plans, activities, assessment tools) that provided evidence of inclusive practices. The teachers created a multimedia presentation at the end of the project that was used as a data source. Similarly, students created documents (e.g., pictures, classroom notes, assignments and projects, videos) that represented their learning. These classroom artifacts were collected by the teachers and were examined by the authors to corroborate research themes. In total, the teachers and students created over 100 artifacts.

3.3.4   Observations

The first author visited the school seven times over a six-month period for a total of 350 min, observing Andy and her Grade 2 students as they engaged in UDL-focused inquiry lessons in their science. The recorded observations were compared with the teacher/student artifacts to determine conformity within data sources.

3.3.5   Field Notes

The first author participated in the six full-day collaborative planning meetings and took notes that were used as a data source to corroborate findings. The group negotiated an agenda for each meeting. The first three meetings focused on examining and reflecting on UDL readings and how UDL had been interpreted by other teachers. During the fourth meeting, after the teachers were comfortable with UDL, they each developed a plan for how to adopt UDL in science. Two collaborative meetings occurred after implementation of UDL principles and guidelines.

3.3.6   Data Analysis

To assist with the process of data analysis and coding, we used qualitative computer software (MAXQDA, version 20.4, 2020) to organize and manage the large amount of data collected. A grounded theory approach (Strauss & Corbin, 1998) was used, and we started by coding all the relevant data to clarify ideas, concepts, and categories. The coding process occurred in three stages. In the first stage, we followed open-coding techniques and read the various sources of data (i.e., interview transcripts and written reflections) multiple times to identify common events or ideas described by teachers involving UDL principles and inclusive practices in science. In the second stage, the authors discussed and compared initial codes; any discrepancies were resolved through discussion. At the third stage, axial coding was employed to assemble initial codes into categories and subcategories, which were reviewed again to compare events and incidents across the data coded. Further, peer debriefing with the teachers and triangulation methods, such as comparing the various data sets and engaging in researcher collaborative reflection, were used to establish trustworthiness. The data analysis process highlighted evidence that supported or contested themes that emerged from the various data sources (Flick, 2018). Thus, this analysis provided us with a thorough and comprehensive understanding of the complex phenomena of teacher professional learning about inclusivity, inclusive practices, and UDL principles in the context of the science teaching and learning from the perspective of this group of participating teachers.

3.4 Adoption of UDL in Science: A Case Vignette

Through a case vignette, we describe the experiences of Andy during planning and implementation of UDL in science, and her thoughts on its impact on student learning. We report outcomes of the study based on Andy's changing views of inclusion and UDL and her perceptions of the value of the UDL framework for all teachers and students. Sources of the data are identified in the subsequent sections as I (interview), R (teacher reflection), A (teacher or student artifact), O (observation), and N (field note).

Andy's Grade 2 class consisted of 9 girls and 13 boys; nine of these children had identified learning needs in terms of social skills, behavioural regulation, and writing and mathematics skills. Andy's beliefs about students at the beginning of her inquiry reflected key tenets of inclusion such as "all learners need to be valued, all have a right to an accessible education, [and] all students need to be productive members of the learning environment" (R). These beliefs were affirmed through the study. In exploring the merit of UDL with her colleagues, Andy hoped to "effectively diminish learning barriers to foster student success, become more fluent with and have a deeper understanding of UDL guidelines as well as the tools to support learning for all, [and] ... help students effectively communicate their learnings and to understand themselves as learners" (I).

3.4.1 Andy's Planning, Implementation, and Impact on Learning

3.4.1.1 Planning

Numerous informal discussions around planning occurred amongst the teachers during lunch periods and after school. The teachers reviewed the literature on UDL using key resources (e.g., Brookes Novak, 2016; Brookes Publishing, 2016; University of Washington, 2020a, 2020b) to inform their understanding. After becoming familiar with the UDL framework and realizing which principles and guidelines could support student learning, Andy selected two key guidelines as a focus for her project:

1. Develop self-assessment and reflection

This guideline considers the capacity of students to engage in self-regulation and to monitor to what degree they are making progress addressing goals and developing independence. Consequently, the teacher needs to introduce "multiple models and scaffolds of different self-assessment techniques so that they [students] can identify, and choose, ones that are optimal" (CAST, 2020b, para. 1).

2. Guide appropriate goal-setting

Using this guideline to inform planning, teachers assist students in developing goal-setting strategies through the introduction of supports such as:

- Provide prompts and scaffolds to estimate effort, resources, and difficulty
- Provide models or examples of the process and product of goal-setting
- Provide guides and checklists for scaffolding goal-setting
- Post goals, objectives, and schedules in an obvious place (CAST, 2020c, para. 1).

Andy then developed two inquiry questions to guide her action research project: "How do I promote student self-assessment and student self-understanding as learners in science?" and "How can I scaffold graduated levels of learning to help learners become more self-guided?"

For her inquiry project, Andy focused on a science unit about relative position and motion. In developing her curriculum with UDL principles and practices, she focused on several prescribed learning outcomes, such as describe the position of an object in terms of change in position relative to other objects, investigate factors that affect movement, communicate using scientific terminology, and work with others in exploring and investigating (see Table 3.1 for the complete list of targeted outcomes). Andy planned a set of learning episodes, each ranging from 1 to 2 hours. The main tools she used in adopting UDL were:

- Conceptual tools: curriculum guide and resources, UDL framework and literature, a UDL lesson-planning guide (see https://www.theudlproject.com/udl-tools---all-grades.html and Appendix B for examples of conceptual tools)
- Physical tools: iPads, iPad applications, interactive white board, coding technology.

Table 3.1 Learning outcomes targeted by Andy in the science unit

Students will be able to:	Nature of outcome
Describe the motion of an object in terms of a change in position relative to other objects	Knowledge
Describe the position of an object relative to other positions or stationary objects	Knowledge
Place an object in an identified position relative to another object or position	Knowledge
Describe the position of objects from different perspectives	Knowledge
Investigate different patterns of movement	Knowledge
Investigate factors that affect movement	Knowledge
Pose questions that lead to exploration and investigation	Skill
Communicate using scientific terminology	Skill
Predict based on an observed pattern	Skill
Work with others in exploring and investigating	Attitude
Willingly observe, question, and explore	Attitude

As Andy examined her curriculum outcomes and considered what UDL would look like in her classroom (i.e., "classroom environment, tools for learning, as well as the presentation of information"), she worked on establishing a "respectful, collaborative environment to foster a communicative, accepting learning space" (R). She started early in the school year helping students monitor their own learning, setting a foundation for self-reflection and goal-setting. Andy described this process during a November planning session:

> I started very, very small in literacy. I expanded it to science eventually so that when I actually got to implement my project, those types of things weren't going to be barriers in and of themselves; they already had a knowledge—a working knowledge and expectations of that. So, basically, at the beginning of each component or lesson, I would make the criteria for a lesson or task very explicit. So, they knew right off the bat that this is what [I am] going to be looking for when they came to do a conference with me, or if we sat and did some self-reflection as a group; those were the goals that I asked them to reflect on, you know, within their writing. So, it just set a foundation that would help me throughout my project implementation.

To inform her practice and to determine whether she was answering her research questions, Andy planned to use a variety of data collection methods and sources, including classroom observational notes, student-generated work, pictures and classroom video, and teacher-written reflections about unfolding events and developing impressions and understandings.

3.4.1.2 Implementation

Andy was excited to start implementation of her science unit on position and motion. She commented: "I have worked to build a goal-setting atmosphere in my classroom and in my literacy curriculum. However, I have never implemented it within science; it will be interesting to see how it impacts science learning" (R). During her introductory class in the science unit, Andy focused on helping students brainstorm a list of common objects that moved in different ways (e.g., ball, toys, slinkies, yo-yo). Students were given the opportunity to "make connections to their personal experiences in order to make the concepts of physical science real to them" (A). In a follow-up the next day, students visited the playground to observe and take pictures of different types of movement (e.g., leaves moving, a bird flying, a flag moving). To document their experiences through pictures and to describe the movement in their own words, students used an iPad application called Seesaw that contains a built-in annotation tool to record their understanding. Andy remarked during a classroom visit that "the Seesaw app proved to be an effective tool for student reflection showing evidence of movement in our world" (O). Throughout the implementation of the unit, the app was used frequently by the students and "provided information to help … [Andy] effectively scaffold lessons" (N).

To help develop student self-assessment and guide goal-setting during different science learning activities, Andy used a number of strategies, including asking students to post *I-wonder* questions on sticky notes on the wall at the beginning of the unit, encouraging students to examine and re-examine the I-wonder questions throughout the unit, posting the objectives of each lesson at the beginning of each class, and guiding whole-class student reflections about what had been learned in previous classes. Andy felt that the strategy of examining and re-examining questions was particularly important "to assist students in recognizing accountability and self-monitoring ... it has been a very effective reflective tool for both the students and me" (R).

Andy and the students explored directionality, patterns of movement, and positions of objects in relation to other objects and from different perspectives. During one class, they reviewed movement and positional language from their word wall (e.g., left, right, forward, away from, closer). This activity was followed by pairs of students using charades to demonstrate how animals move. Some students generated new I-wonder questions about animal movement (e.g., "I wonder how worms dig through the ground" [N]). A range of other learning activities were incorporated into the unit such as engaging in unplugged coding activities, reading children's literature, playing a game of Simon Says, sorting objects based on the types of movement, exploring ways to get objects to move, investigating ramps to determine the impact of varying ramp height on the speed of a toy car and distances travelled, and manipulating perspectives by rolling a cube with position-related terms and asking students to describe the position of an object in the classroom in relation to a stationary object.

In one activity, later in the unit, students were given the opportunity to engage in a design challenge. As a whole class, they developed criteria: make a small toy car move through two right turns, two left turns, go over an object, and go under an object. The challenge was to design a track that would allow the car to meet the developed criteria. Students worked in groups of three to design their car tracks using materials of their choice, such as cardboard, glue, and tape. Andy reported that many of the children had difficulty with the task initially, but that the activity was a success:

> Overall, 21–22 students were engaged in this activity. However, self-monitoring skills were more difficult during this activity. Many did not apply their knowledge from previous lessons. The challenge took a long time, and we cleaned up to continue the next day. Upon returning the next day, we displayed the projects and discussed the components of the criteria that were actually met. I asked them to test their creations and redesign where they felt it was needed. This reflection activity provided an opportunity to evaluate and "fix up" any sections that did not meet criteria. After the second day, I was quite pleased with the results of assessing their projects! All groups were successful at varying degrees. (R)

In summarizing her planning framework for the unit and implementation, Andy reviewed her approach to fostering student reflection, self-monitoring, and goal-setting:

> I have attempted to scaffold lessons which build the foundation for all students to succeed.... This UDL research cycle has been a very thought-provoking journey. I have investigated areas of multiple means of engagement and multiple means of action and expression. There exists much crossover between these principles within the framework. (I).

Andy expressed concerns about one student who was identified as having behavioural challenges and the "inability to contribute meaningfully within a peer group (social skills)" (I). While there was a heavy emphasis placed on student collaboration throughout the unit, Andy recognized that working in collaborative groups was a barrier to this student's learning. She further reflected on this concern at a collaborative meeting:

> This continues to be a point of reflection for me as an educator. In response to these social/behavioural needs, I will continue to limit the groupings to partners so she will have more success during the activities and explorations. I will also pair her with a positive role model to aid in collaboration and scaffolding. In this way, she is exposed to effective collaboration in the most positive social situation possible. UDL has helped me be more reflective of her learning barriers, both academically as well as behaviorally. (N)

She reflected on the need to keep reviewing her learner profiles, "attempting to predict barriers to learning and implement a plan for scaffolding lessons and concepts" (I).

Andy's primary pedagogical strategies for scaffolding student complex tasks were teacher modelling and probing questioning, small- and whole-group discussions, ongoing student written reflections, and monitoring group interactions. By the end of the unit, Andy commented that her students "learned how to relate the activities to the established goals from the beginning of the lesson, making the learning process more cyclic" (I). For Andy, UDL "goes much deeper than providing for learner needs with the expectation they will acquire a specific set of knowledge/skills; it is a framework that fosters an awareness of themselves as learners in a goal-directed, supported learning environment" (N). Thus, an important part of Andy's inquiry was to determine the impact on her students' learning.

3.4.1.3 Impact on Learning

In analyzing her data and with support from her colleagues who assumed the role of critical friends, Andy concluded that consistent implementation of UDL principles did enable her students to become more self-directed learners. "Through the implementation of UDL in planning and preparation for unit delivery, students were more aware of their personal understandings and could, more successfully, articulate science terms, results of explorations, and new wonderings" (I). Moreover, Andy "noted that most students were able to more effectively communicate their understanding and reasoning in science. They were invested in their learning, understood they were accountable for learning … and enjoyed sharing their thoughts and opinions as well as their new questions" (R). While all students progressed, Andy felt some students needed more time and additional opportunities to become goal-oriented and self-reflective at the level she had expected. Having taught Grade 2 for several years, Andy felt comfortable with the science curriculum. However, she reported that this cycle of action research allowed her "to examine the science content through a different lens,… empowering students to become 'expert learners' in their own right—to be able to access and achieve goals by making deliberate learning choices" (N). When asked if she would continue to use the UDL framework as a guide, Andy replied that she hoped to "go further with … communication and being able to allow them choice; to continue in science and to extend to math" (N). By adopting UDL principles in her classroom, Andy helped her students become more aware of their own learning through improved communication, self-reflection, and goal-setting. In this way, these students had more access to the science curriculum by making optimal choices about their own learning.

3.4.2 Andy's Changing Views of Inclusion and UDL

At the beginning of this study, the UDL framework was not commonly used in Newfoundland and Labrador schools and had only been introduced on a provincial level in the previous year. While they were very comfortable with the concept of DI, the participating teachers wanted to explore the potential of UDL for "empowering their students" and "making the curriculum accessible to everybody" (N). Having become comfortable with inquiry-based and design approaches to teaching and learning through their previous action research cycles, they thought this cycle was an opportune time to examine their perspectives and practices more closely in relation to inclusion using UDL.

In exploring UDL, the teachers worked closely with the first author by reviewing literature and exploring emerging ideas together. At the end of the first collaborative planning meeting, the teachers reported feeling overwhelmed by the framework as they considered the three broad UDL principles and their corresponding 31 guidelines and checkpoints. Later, in the fall, after more reading and discussion, Andy reflected on the nature of the UDL framework:

> My knowledge and understanding of the UDL framework are ever emerging and becoming deeper with each article and piece of information as I sort through my preconceived notions with the facts of this framework. We have been immersed in an inclusive approach for education, equipped with the theories of differentiated instruction and assessment. I feel that it is only today that I truly understand the difference between the two. (R)

Distinguishing between UDL and DI created dissonance for Andy. Initially, she thought "this was just a new word for DI" (N). However, as they continued to delve into the area, she noted key differences between the two approaches: "UDL has a different lens. Instead of me making those decisions for the students, they are making decisions for themselves and what they need as learners as long as I'm scaffolding the lessons for them" (I). She recognized that adopting a UDL perspective requires being proactive and examining the curriculum, assessment, and students' learning needs prior to planning instruction.

3.4.3 Andy's Perspectives on the Value of the UDL Framework

In adopting the UDL framework as a lens to inform curriculum planning and class-room practice, Andy was able to enhance her own professional understanding of how to support all students in the classroom. She reported becoming more knowledgeable about the curriculum and engaged in more intensive reflection about how to support student learning. Andy said that one of the most important insights she developed was in terms of empowerment of students:

> The UDL lens empowers students to become 'expert learners' in their own right—to be able to access and achieve goals by making deliberate learning choices. Therefore, learning is fostered through personal choices rather than the teacher making choices for the learner. (I)

When asked to comment on the value of the UDL framework overall for creating inclusive learning environments, Andy suggested that teachers should start "embed-ding [the principles] into the curriculum early … so they [students] become accus-tomed to it and with each tiny step, it becomes easier for students" (I). The group felt strongly that all teachers should apply UDL, with Susan commenting that its use across the curriculum could support "students to become more strategic learners and … have choice and voice in making decisions about the best materials, methods, and tools for both learning/exploring as well as communicating" (R). Furthermore, Andy suggested that "student learning is impacted by barriers, which should be predicted during the intentional planning process. Using the UDL framework … enables educators to facilitate student independence and goal-directed learning" (I).

The teachers acknowledged the importance of having support through profes-sional learning premised on collaboration, inquiry, and relevance when adopting the UDL framework. Susan stated, "I think teachers need opportunities for collaboration in a professional learning community. Teachers need to work toward shared goals … reflection and collaboration are needed…. [We need to] explore new ideas,… test them, pose questions, get answers, and put them in action" (N). Andy felt strongly that professional learning should be driven by the goal of "supporting student learning strategically and deliberately" (R).

3.5 Conclusions and Recommendations

Andy and her group members found that UDL was a useful approach for creating inclusive classrooms in science. A substantial body of research indicates that teachers struggle with creating inclusive learning environments (Avramidis & Norwich, 2002; Damianidou & Phtiaka, 2018; Florian & Graham, 2014; Sagner-Tapia, 2018; Southerland et al., 2011). Villanueva et al. (2012) suggested that an obstacle causing this struggle is that "instructional adaptations and inclusion appear to be a consid-erable task for which science teachers are ill-equipped to undertake" (p. 189). Barriers to inclusive instruction of K–12 teachers identified by Southerland et al.

(2011) include teacher beliefs about learner characteristics, teaching infrastructure such as curriculum outcome expectations, and personal beliefs such as self-efficacy relating to teaching knowledge and skills. Yet, some of the earliest studies of diverse learners demonstrated that students with learning disabilities benefit from inquiry-based instruction (Mastropieri et al., 1997; Scruggs & Mastropieri, 1994). Lee and Picanco (2013) theorized that "differentiated instruction, UDL, and co-teaching can be used effectively in concert with planning for the phases of learning to create optimal learning experiences for students" (p. 143). Using evidence from teaching and reflecting on student experiences, the teachers in the case study presented here demonstrated that UDL is a viable framework for designing inquiry lessons and one way to create more inclusive science classrooms.

We strongly encourage teachers to adopt the UDL framework as a means to create inclusive science education. UDL provides teachers with a comprehensive guide to consider many aspects of their professional practice in relation to making the curriculum accessible to all learners. The teachers in this study started the inquiry with a collective examination of the UDL framework and then identified one or more guidelines within the three UDL principles to guide planning and teaching that support inclusive student learning. For example, Andy wanted her students to increase their self-guided learning and gradually establish a culture of student self-reflection and goal-setting. Using the curriculum and specific UDL guidelines as conceptual tools and in concert with physical tools (e.g., technology and science materials), she scaffolded students' self-monitoring during inquiry-based learning. In reviewing learner profiles, Andy found that the UDL framework fostered student self-awareness. The *bite-sized* focus on UDL principles by Andy was effective for enhancing her instruction and being more responsive to all students' needs. Thus, the authors suggest that teachers start small when adopting UDL since the framework, with its three broad principles and 31 guidelines, may be perceived as overwhelming.

Based on the experiences of Andy, this case study provides evidence that inquiry-based science instruction may be used to embed UDL principles into the curriculum, thus helping to create inclusive classrooms. Watt et al. (2013) found that the activity level of inquiry-based lessons allowed teachers to address the various needs of diverse students by implementing a range of supports, which aligns with the UDL framework. A Canadian meta-analysis of STEM learning challenges experienced by students with learning disabilities aligns with the use of UDL principles by the teachers in this study, which recommended that "presenting similar information using multiple representations can help students to access and process pertinent information" (Asghar et al., 2017, p. 244). Stavroussi et al. (2010) suggested that learning strategies in science education that emphasize hands-on activities and real-life experiences can support learning for students with a mild to moderate intellectual disability. Likewise in this study, Andy found that using the principles of UDL to intentionally plan science experiences for inclusion resulted in increased success for all students. The participating teachers' collaborative review of Andy's action research data suggested that her students employed their new UDL metacognitive skills, thus addressing motion-related curriculum outcomes and relating the activities to their personal learning goals. The science inquiry lessons that Andy designed using selected UDL principles empowered the students and fostered teacher responsiveness by encouraging student self-reflection and goal-setting.

Finally, we strongly encourage teachers to engage in collaborative professional learning to address some of the potential barriers to inclusive instruction (Asghar et al., 2017; Southerland et al., 2011). The teachers in this study acknowledged the importance of mutual support in guiding their planning, teaching, reflections, and professional learning. At times, they struggled with instructional decisions such as the use of differentiated instruction versus UDL principles. Ultimately, they chose strategies and approaches that empowered students to make personal learning choices during science inquiry and fostered learning that was accessible to all. Thus, the application of the UDL framework to a science unit on motion promoted the development of an inclusive student-centered learning environment, helping students become more self-directed and reflective as they addressed a range of learning outcomes.

References

Aldani, G. (2020). Are we ready for inclusion? Teachers' perceived self-efficacy for inclusive education in Saudi Arabia. *International Journal of Disability, Development and Education, 67*(2), 182–193. https://doi.org/10.1080/1034912X.2019.1634795

Asghar, A., Sladeczek, I. E., Beaudoin, E., & Mercier, J. (2017). Learning in science, technology, engineering, and mathematics: Supporting students with learning disabilities. *Canadian Psychology, 58*(3), 238–249. https://doi.org/10.1037/cap0000111

Avramidis, E., & Norwich, B. (2002). Teachers' attitudes towards integration/inclusion: A review of the literature. *European Journal of Special Needs Education, 17*(2), 129–147. https://doi.org/10.1080/08856250210129056

Basham, J. D., & Marino, M. T. (2013). Understanding STEM education and supporting students through universal design for learning. *Teaching Exceptional Children, 45*(4), 8–15. https://doi.org/10.1177/004005991304500401

Blanton, L. P., Pugach, M. C., & Florian, L. (2011). *Preparing general education teachers to improve outcomes for students with disabilities.* American Association of Colleges for Teacher Education; National Center for Learning Disabilities. https://www.ncld.org/wp-content/uploads/2014/11/aacte_ncld_recommendation.pdf

Brookes Publishing. (2016, February 11). *Use UDL in your lesson planning to enhance your teaching.* [Video]. https://www.youtube.com/watch?v=B5JWvCaXk-8

Campbell, C., Osmond-Johnson, P., Faubert, B., Zeichner, K., Hobbs-Johnson, A., Brown, S., & Steffensen, K. (2016). *The state of educators' professional learning in Canada: Final research report.* Learning Forward. https://learningforward.org/wpcontent/uploads/2017/08/state-of-educators-professional-learning-in-canada.pdf

Center for Applied Special Technology. (2018). *Universal Design for Learning Guidelines version 2.2.* https://udlguidelines.cast.org/

Center for Applied Special Technology (CAST). (2020a). *About universal design for learning.* https://udlguidelines.cast.org/engagement/?utm_source=castsite&utm_medium=web&utm_campaign=none&utm_content=aboutudl

Center for Applied Special Technology. (2020b). *Develop self-assessment and reflection.* https://udlguidelines.cast.org/engagement/self-regulation/self-assessment-reflection

Center for Applied Special Technology. (2020c). *Guide appropriate goal-setting.* https://udlguidelines.cast.org/action-expression/executive-functions/goal-setting/goal-setting

Collins, A., Philpott, D., Fushell, M., & Wakeham, M. (2017). *Now is the time: The next chapter in education in Newfoundland and Labrador.* Premier's Task Force on Educational Outcomes. https://www.gov.nl.ca/education/files/task_force_report.pdf

Damianidou, E., & Phtiaka, H. (2018). Implementing inclusion in disabling settings: The role of teachers' attitudes and practices. *International Journal of Inclusive Education, 22*(10), 1078–1092. https://doi.org/10.1080/13603116.2017.1415381

Dymond, S. K., Renzaglia, A., Rosenstein, A., Chun, E. J., Banks, R. A., Niswander, V., & Gibson, C. L. (2006). Using a participatory action research approach to create a universally designed inclusive high school science course: A case study. *Research & Practice for Persons with Severe Disabilities, 31*(4), 293–308. https://doi.org/10.1177/154079690603100403

Farrell, P. (2016). Promoting inclusive education in India: A framework for research and practice. *Journal of the Indian Academy of Applied Psychology, 42*(1), 18–29. http://www.jiaap.org.in/Sample.aspx?Sub=JIAAP%20January%202016

Finnegan, L. A., & Dieker, L. A. (2019). Universal design for learning-representation and science content: A pathway to expanding knowledge, understanding, and written explanations. *Science Activities, 56*(1), 11–18. https://www.tandfonline.com/doi/abs/10.1080/00368121.2019.1638745

Flick, U. (2018). Triangulation in data collection. In U. Flick (Ed.), *The SAGE handbook of qualitative data collection* (pp. 527–544). Sage.

Florian, L., & Graham, A. (2014). Can an expanded interpretation of phronesis support teacher professional development for inclusion? *Cambridge Journal of Education, 44*(4), 465–478. https://doi.org/10.1080/0305764X.2014.960910

International Disability and Development Consortium. (2017). *Call to action to invest in disability-inclusive education.* https://www.iddcconsortium.net/blog/call-to-action-to-invest-in-disability-inclusive-education/

Kemmis, S., & McTaggart, R. (2005). Participatory action research: Communicative action and the public sphere. In N. K. Denzin & Y. S. Lincoln (Eds.), *The Sage handbook of qualitative research* (pp. 559–603). Sage.

Krischler, M., Powell, J. J. W., & Pit-Ten Cate, I. M. (2019). What is meant by inclusion? On the effects of different definitions on attitudes toward inclusive education. *European Journal of Special Needs Education, 34*(5), 632–648. https://doi.org/10.1080/08856257.2019.1580837

Lambe, J., & Bones, R. (2006). Student teachers' perceptions about inclusive classroom teaching in Northern Ireland prior to teaching practice experience. *European Journal of Special Needs Education, 21*(2), 167–186. https://doi.org/10.1080/08856250600600828

Lee, C., & Picanco, K. E. (2013). Accommodating diversity by analyzing practices of teaching (ADAPT). *Teacher Education & Special Education, 36*(2), 132–144. https://doi.org/10.1177/0888406413483327

Leifler, E. (2020). Teachers' capacity to create inclusive learning environments. *International Journal for Lesson and Learning Studies, 9*(3), 221–244. https://doi.org/10.1108/IJLLS-01-2020-0003

Marino, M. T. (2010). Defining a technology research agenda for elementary and secondary students with learning and other high-incidence disabilities in inclusive science classrooms. *Journal of Special Education Technology, 25*(1), 1–27. https://doi.org/10.1177/016264341002500101

Mastropieri, M. A., Scruggs, T. E., & Butcher, K. (1997). How effective is inquiry learning for students with mild disabilities? *Journal of Special Education, 31*(2), 199–211. https://doi.org/10.1177/002246699703100203

Materechera, E. K. (2020). Inclusive education: Why it poses a dilemma to some teachers. *International Journal of Inclusive Education, 24*(7), 771–786. https://doi.org/10.1080/13603116.2018.1492640

Merriam, S. B. (1998). *Qualitative research and case study applications in education.* Jossey-Bass.

Merriam, S. B., & Tisdell, E. J. (2015). *Qualitative research: A guide to design and implementation.* John Wiley & Sons.

Meyer, A., Rose, D. H., & Gordon, D. (2014). *Universal design for learning: Theory and practice.* CAST Professional Publishing.

Novak, K. (2016). *UDL now!: A teacher's guide to applying universal design for learning in today's classrooms.* CAST Professional Publishing.

Opertti, R., Brady, J., & Duncombe, L. (2009). Moving forward: Inclusive education as the core of education for all. *Prospects, 39*(3), 205–214. https://doi.org/10.1007/s11125-0099112-3

Rappolt-Schlichtmann, G., Daley, S. G., Lim, S., Lapinski, S., Robinson, K. H., & Johnson, M. (2013). Universal design for learning and elementary school science: Exploring the efficacy, use, and perceptions of a web-based science notebook. *Journal of Educational Psychology, 105*(4), 1210–1225. https://doi.org/10.1037/a0033217

Riedell, K. E. (2016). *Understanding curriculum, instruction and assessment within eighth grade science classrooms for special needs students.* [Doctoral dissertation, California State University, Los Angeles]. UCLA Electronic Theses and Dissertations. https://escholarship.org/uc/item/3bv393j5

Sagner-Tapia, J. (2018). An analysis of alterity in teachers' inclusive pedagogical practices. *International Journal of Inclusive Education, 22*(4), 375–390. https://doi.org/10.1080/13603116.2017.1370735

Saloviita, T., & Schaffus, T. (2016). Teacher attitudes towards inclusive education in Finland and Brandenburg, Germany and the issue of extra work. *European Journal of Special Needs Education, 31*(4), 458–471. https://doi.org/10.1080/08856257.2016.1194569

Scruggs, T. E., & Mastropieri, M. A. (1994). The construction of scientific knowledge by students with mild disabilities. *Journal of Special Education, 28*(3), 307–321. https://doi.org/10.1177/002246699402800306

Shore, L. M., Randel, A. E., Chung, B. G., Dean, M. A., Holcombe Ehrhart, K., & Singh, G. (2011). Inclusion and diversity in work groups: A review and model for future research. *Journal of Management, 37*(4), 1262–1289. https://doi.org/10.1177/0149206310385943

Southerland, S., Gallard, A., & Callihan, L. (2011). Examining teachers' hurdles to 'science for all.' *International Journal of Science Education, 33*(16), 2183–2213. https://doi.org/10.1080/09500693.2010.530698

Stavroussi, P., Papalexopoulos, P. F., & Vavougios, D. (2010). Science education and students with intellectual disability: Teaching approaches and implications. *Problems of Education in the 21st Century, 19*(1), 103–112. http://www.scientiasocialis.lt/pec/node/files/pdf/vol19/103-112.Stavroussi_Vol.19.pdf

Strauss, A., & Corbin, J. (1998). *Basics of qualitative research: Techniques and procedures for developing grounded theory.* Sage.

Stronge, J. H., Ward, T. J., Tucker, P. D., & Hindman, J. L. (2007). What is the relationship between teacher quality and student achievement? An exploratory study. *Journal of Personnel Evaluation in Education, 20*(3–4), 165–184. https://doi.org/10.1007/s11092-008-9053-z

United Nations Educational, Scientific and Cultural Organization. (2000). *Education for all 2000 assessment: Global synthesis.*

United Nations Educational, Scientific and Cultural Organization. (2005). *Guidelines for inclusion: Ensuring access to education for all.*

University of Washington. (2020a). National Center for Universal Design for Learning. https://www.washington.edu/doit/national-center-universal-design-learning.

University of Washington. (2020b). *National Center for Universal Design for Learning.* https://www.washington.edu/doit/national-center-universal-design-learning

Villanueva, M. G., Taylor, J., Therrien, W. J., & Hand, B. (2012). Science education for students with special needs. *Studies in Science Education, 48*(2), 187–215. https://doi.org/10.1080/14703297.2012.737117

Watt, S. J., Therrien, W. J., Kaldenberg, E., & Taylor, J. (2013). Promoting inclusive practices in inquiry-based science classrooms. *Teaching Exceptional Children, 45*(4), 40–48. https://doi.org/10.1177/004005991304500405

Wilson, V. A. (2014). *Secondary general education teachers' attitudes toward inclusion.* [Doctoral dissertation, Regent University]. ProQuest Dissertations Publishing.

Yin, R. K. (1994). *Case study research: Design and methods* (2nd ed.). Sage.

Karen Goodnough BSc, BEd, MEd (Memorial University of Newfoundland), PhD (Ontario Institute for Studies in Education of the University of Toronto), is a professor of science education and the dean of the Faculty of Education at Memorial University of Newfoundland. She is actively engaged in research that focuses on collaborative action research, preservice teacher education, problem-based learning, science teaching and learning, self-study in teacher education, and teacher learning in STEM (Science, Technology, Engineering, and Mathematics) education. She is a former high school science teacher and spent several years working in the area of gifted education. Her most recent research funding from the Social Sciences and Humanities Research Council (SSHRC) focuses on examining the perspectives and practices of Canadian science teacher educators.

Saiqa Azam BSc and MSc (University of Punjab), BEd (Allama Iqbal University), MEd (Monash University), PhD (University of Calgary), is an associate professor of science education in the Faculty of Education at the Memorial University of Newfoundland. She completed her undergraduate studies in Pakistan and taught there as a science teacher and a science teacher educator. Her research interests include science teachers' pedagogical content knowledge (PCK) and the development of science and STEM teaching identity. She is also interested in equity issues in science education and focuses on the questions of designing inclusive science education for diverse student populations. Current projects focus on studying and documenting science teachers' PCK of teaching science topics, science teachers' PCK related to socio-scientific issues, preservice elementary teachers' science and STEM teaching identity, and inclusive perspectives and practices of science teacher educators.

Patrick Wells BSc (Saint Mary's University), BEd (Dalhousie University), MSc (Dalhousie University), is a PhD candidate (Memorial University of Newfoundland) and a retired science teacher (28 years, high school science). In the last 15 years of his teaching career he focused on action research, ocean education, and leading professional learning with science teachers provincially, nationally and internationally. His research interests include student inquiry in science, science teacher learning, ocean literacy, and lesson study as a model professional development.

Chapter 4
Science in the Spotlight: What Are Monsters Made of? (A Performative Inquiry)

Lynn Fels and Karen Meyer

4.1 Prologue: Performing Science

Out of the slate gray of this rainy Vancouver morning comes a sinuous line of children…. Like a stripe of ants following a collective purpose, they and their teachers move through the university building to the theatre doors. Then, inside the playhouse, kinetic bodies burst from raincoats. The youngest find places in the front rows, staring at the closed curtains, their boots dangling. Back and forth measures time waiting. Older children chat.

"I'm really glad this is science," one whispers to another.

Behind the curtain, ten university students and their two professors attend to last minute details, the performance imagined in science class now just moments from show time.

"Is my make-up okay?" queries one of the Jesters. Einstein and Wendy rehearse their lines in a secluded corner. Einstein's rabbit ears tremble with stage fright.

"Two minutes to curtain time," cautions the director. In the booth, the sound technician cues the CD player. The lighting technician dims the house lights.

The Monster whispers, "Break a leg."

And the play begins. (Meyer & Fels, 1998, p. 22)

More than 25 years ago, we co-taught several science elementary methods courses for preservice teachers in the University of British Columbia Faculty of Education. At the time, Karen was a science educator experienced in developing interactive exhibits

L. Fels (✉)
Simon Fraser University, Burnaby, BC, Canada
e-mail: lynn_fels@sfu.ca

K. Meyer
University of British Columbia, Vancouver, BC, Canada
e-mail: karen.meyer@ubc.ca

© The Author(s), under exclusive license to Springer Nature Switzerland AG 2023
C. D. Tippett and T. M. Milford (eds.), *Exploring Elementary Science Teaching and Learning in Canada*, Contemporary Trends and Issues in Science Education 53, https://doi.org/10.1007/978-3-031-23936-6_4

at a science museum; and Lynn was a doctoral student engaged in conceptualising and articulating drama and theatre as performative inquiry; both of us joined our curiosity and practice around the interplay between science education and drama as performance. Before and after classes, we thrashed out ways to incite participation among students who had modest backgrounds and often little interest in science. We tackled science concepts as everyday physical phenomena by regularly leaving the classroom to become a pendulum on park swings, make billiard shots using angles of reflection, or feel friction at our feet on the ice rink.

We met by coincidence at a noon-hour talk on complexity theory, which was then edging its way into education. Lynn was excited to explore the integration of drama and story into science education. She had been inspired by her work as an artist in education and the play she had co-created with Grade 3 students that explored air pressure. *What the heck is air pressure,* she wondered, *and who cares*? Her experience creating a play with students about air pressure (i.e., history of flight, wind instruments, weather, building and flying a plane on stage) led to her thinking about re-imagining science education and the learning that happens through play. Karen, as a new faculty member, was looking for someone to "play with"; and so upon our first encounter, we recognized kindred spirits.

As the course instructors, perhaps considered by some to be an odd couple, we took a performative and experiential approach to learning how the physical world around us works by integrating science and drama and by creating learning experiences for our science education students. We were curious to explore the phenomena of light, sound, and movement with our students in collaborative, exploratory ways. Our mission became creating provocative situations or conditions as events that would embolden active participation and attention to what happens and what matters—noticing tugs on the sleeve and bringing ideas into presence. Like scientists and performative inquirers, we began with a question: *What if?* that was followed by *What matters? What happens? So what? Who cares?* We engaged in our work through curiosity and play that resulted in a simple question: What happens if you integrate drama and story into science education?

As colleagues, we shared an interdisciplinary, performative, and experiential event in collaboration with our teacher education students and the elementary school teachers and their students from the local school. This pedagogical encounter—embodying the germinating seeds of our emergent scholarship alongside colleagues—remains relevant in today's science education. At that time, we were on the cusp of articulating performative inquiry (Fels, 1999, 2010, 2012) and reconceptualizing science inquiry (Erickson & Meyer, 1997; Meyer, 2000). Ours was an emergent curriculum (Davis et al., 2008; Fels & Meyer, 1997) that was unfolding within the understanding that learning is social, interactive, and collaborative (Christiansen et al., 1997; Clark et al., 1996), leading to transformation of practice and relationship (Goulet et al., 2003). We drew upon theories of enactivism

(Maturana & Varela, 1992; Varela et al., 1993) and complexity (Davis et al., 1996; Waldrop, 1992) that were unexpected strangers crossing the threshold into education. We were curious new colleagues working together to explore what was possible in the liminal space of encounter between science and drama, inquiry, and performance.

This is our story, conceivably historical, of what happened when we created a play with preservice teachers in an elective science education course and the arrival of the script, *Light, Sound, Movin' Around: What are monsters made of?* We extend an invitation to reimagine science education as a performative inquiry of phenomena in a playful encounter and offer both the story and the play—the actual script unburied from the bottom drawer of a filing cabinet—in the retirement of two long careers. Here, we share with you what was one of our most eventful, memorable experiences of teaching science in the hope that you might be inspired to imagine science into play.

4.2 In the Beginning ...

That first day of class in our third year of teaching together, we walked into the classroom with our freshly photocopied course syllabus in hand. Ten students sat with varying degrees of welcome on their faces. When asked why they were taking the course, many admitted to their minimal experience with science and a few brave ones confessed to disliking science. "It's why I have a BA in literature," muttered one student. "We HAVE to teach science," they explained, "we're going to be elementary teachers. We have to teach EVERYTHING!" Once more, we faced a challenge.

Casting worried glances at each other, we set our course syllabus aside with an offer of negotiation. How could scientific concepts become engaging, accessible, and relevant? How, we wondered together, might we engage an exploratory play of science? In a slow surge of ideas, a lone dare was offered. "Let's create a play!" We looked at each other in surprise: *In science class?*

As soon as students understood that they could choose how they would participate, they leapt to action. Three, including the English literature major, volunteered to write the script. Through brainstorming, our play's characters morphed into life and became the heart of our adventure together: Wendy, a ten-year-old girl who likes to figure things out by doing; Einstein, her stuffed pink bunny with an encyclopaedic memory; two non-stop talking jesters; a cool keyboard player; a prop person (to do the heavy lifting); and a villain, of course, the Shadow Monster. A few of us had technical skills and offered to work behind the scenes; a few others chose to act; as a team, we choreographed action, music, sound effects, lighting, and set changes on stage.

4.2.1 Exploring and Problem Solving

The initial script outline served as a catalyst for our science explorations. Two-word questions guided our participation:

- What matters? (variables)
- What if? (manipulation)
- What happens? (results)
- Who cares? (impact).

Attending to the science within our play called forth the scientists within us. However, the script posed several problems that the students had to resolve. How do we change a pile of clothes into the Shadow Monster? How do we create Einstein's coloured shadows? How do we make the different sounds of a rabbit chewing a carrot? How will we use science to help Einstein and Wendy defeat the Shadow Monster?

Solving shadow puzzles became a critical part of the course curriculum and experimental activity in the science laboratory. None of us, for example, considered presenting the monster in any form other than an actual shadow performing on stage, able to move around and change size. We could have dressed an actor as a shadow or played a prerecorded video of a shadow. But no, such possibilities were out of the question: the English literature major insisted that we have a real Shadow Monster!

Our stage settings and special effects started as simulations in the laboratory. There, materials prompted ideas (What if?) and possibilities were tested (What happens?). According to the story, the Shadow Monster needed to appear big, real, and scary. The script called for coloured shadows to surprise even Einstein. And when Einstein flowed down a drain on stage, we had already worked out which direction and why, by turning on the tap in the sink (clockwise, of course).

After experimentation and trials in the science laboratory, we travelled across campus to the theatre where a movie screen on the stage became a key element in these phenomena of shadow formation. We held our breath as one of us climbed the 25-foot ladder to slip in the coloured gels in the theatre lights. Would our laboratory shadow experiments work on stage? Would our scary Shadow Monster perform on cue?

4.2.2 Pedagogical Opportunities

In tandem, our students visited the elementary school of Grades K–7 nearby, within walking distance to the university. As a science educator, Karen had an ongoing relationship teaching science in the school. Previously, for example, she had explored the physics of swinging in the playground with students and their teacher. Working in small groups, our preservice teachers introduced the children to the phenomena that they would encounter in our upcoming play as Wendy and Einstein travelled through the lands of light, sound, moving around. Several afternoons, we hauled the university's light equipment to their classrooms to apply our learning with the children. With us, children experimented manipulating their own shadows, moving toward and away from the lights, and they learned how to create shadows of different colours.

Our decision to include local elementary school children ensured pedagogical opportunities. The preservice teachers gained valuable teaching experience, by learning to let their young charges learn by doing. The children gained relevant scientific information connecting them to the play they would later see in the theatre: "I'm really glad this is science." We surrendered control of the course syllabus to play and learn inside a collaborative space co-created with our students.

Our class visits inspired excited recognition during the play as the children spotted familiar shadows in colour and understood the science behind defeating the Shadow Monster. Our work in the classrooms before the play unlocked mysteries in the story. This unconventional approach to theatre—giving away secrets beforehand—allowed Einstein to engage the children in the audience in pedagogic play on stage. Stuffy Einstein's line *There's a scientific reason …* became a cue and a clue for children to shout out from their seats what their pink six-foot friend should do next. Happily for the science educators within us, they knew why.

4.2.3 Science as Performative Inquiry

The play became a performative inquiry, an action site of improvisation, innovation, and learning. For example, in our first rehearsal, we debated how to show shadow clothes landing on a shadow chair on the screen when Wendy cleaned up and tossed her clothes offstage. Would what we had imagined in the science laboratory work on stage? The Prop Person, standing in the wings, threw a shirt, a pair of pants, another shirt, timing the movements precisely, onto a chair behind the movie screen. The shadows of clothes landed on a shadow chair projected by a light placed on the stage floor behind the chair. These shadows turned into our Monster Shadow projected on the screen. We cheered! We celebrated our collective creativity and teamwork that made the moment possible.

Phenomena shaped our engagement even as we shaped the phenomena: the script, like a science experiment, was subject to conditions (What matters?) that required brainstorming, improvisation, trial and error. For example, late into rehearsals Scene

Four had no ending, no direction, no exit, with Einstein stuck on stage. The scene's incompleteness caused consternation as we neared the performance deadline. Yet, we tenaciously persisted, collectively tackling the problem of luring the water-shy Einstein to walk across a body of water (made from light) on a log in the scene's world of movement.

"Could Einstein be following the smell of carrots?" Someone asked.

"Impossible! He has nightmares about carrots!" The Director rebutted.

"He's terrified of water! What's his motivation?"

"I've got it! Broccoli."

"And what if the broccoli is on a fishing rod? I could be off stage." The Prop Person proposed.

"Yes!" We all agreed enthusiastically.

"I love broccoli," confided Einstein to the Keyboard Player.

Similarly, the creation and defeat (for now) of the Shadow Monster arrived through trial and error as Einstein and Wendy moved into action through worlds of sound, light, and shadows cast on a movie screen. And when the Shadow Monster rose up from the chair, the proverbial monster under the bed sent chills down our spines.

We designed our special effects and special-effect-performers with audience participation as one of our primary goals. The performance of the script actively engaged the children: creating a rainstorm, playing a game show on sound, shouting warnings to Einstein. In one scene, for example, the children called out what Einstein needed to do to create a shadow big enough to scare himself. "Back up!" the children loudly directed in teaching Einstein how to manipulate the size of his shadow. Of course, they knew first-hand about shadow formation as "It depends …"—where the object is in relationship to the light source. The script, with its characters and actions, performed physical science phenomena we had all manipulated in the laboratory and explored together in the children's classrooms. Phenomena and performance merged into pedagogy.

In theatre, the fourth wall is an imaginary wall located between the audience and the actors. Performers are perceived by the audience to be contained and interacting within four walls of an imaginary world as if the audience is not there. To break the fourth wall, an actor might slip an aside or word of warning to the audience or invite audience members to interact with the actors on stage (Quinlan & Duggleby, 2009). As Einstein, Wendy, and our energetic jesters engaged the children in our play, we as educators broke the fourth wall—the wall between disciplines, between precision and creativity, between audience and actors, between expertise and exploration, and between students and teachers. Drama experimented in the laboratory. Science improvised on stage. The real scientist, Albert Einstein (1929), said, "I am enough of the artist to draw freely upon my imagination. Imagination is more important than knowledge. Knowledge is limited. Imagination encircles the world" (p. 117).

The curricular curtain between science and drama became transparent like a theatre scrim (i.e., a semi-transparent curtain that appears opaque until lit from behind). We created a new monster: a pedagogical play—experiencing with our students science through drama and drama through science. When we negotiated the course syllabus, we had not yet imagined this emergent curricular world where Einstein and Wendy would co-exist with our learning goals, nor had we foreseen mischievous jesters in play with 200 children creating a rainstorm by clapping their hands and stomping their feet. The collective creativity of our students, with Wendy and Einstein's help, chased away the Shadow Monster to the cheers of our young audience.

Our story illustrates the pedagogical possibilities when we invited our science-hesitant education students to use their imaginations and skills to problem solve, playfully engage with science, and defeat the shadow monsters of not knowing, of rote memorized learning, and of scientific bafflegab that plagued poor Einstein, thus encouraging students of diverse interests to learn how to become science teachers in action. We offer our play as an example of what is possible when play, imagination, curiosity, and creative action enter a preservice science teacher education classroom—and how creating a play and making shadows in the science laboratory and elementary classrooms led to an engaged audience of school children, calling out to Einstein, telling the hapless rabbit what to do, revealing what they knew that Einstein did not. These scientists in the making had learned the art of enlarging and shrinking shadows of objects with our preservice teachers in their classroom.

4.3   The Play

Back in the theatre auditorium, sitting in the audience, we smile at each other, nervous smiles, waiting for science to unfold in a play of light, sound, and moving around. The auditorium lights dim. Two jesters shyly, slyly, arrive. A disturbance in the back entrance. You—our reader—your ticket in hand, scurry down the aisle, peering through the darkness, looking for somewhere to sit. *Psssst! Over here! We've saved a seat for you. You've arrived just in time! Come, sit with us!*

Light, Sound, Movin' Around! What are Monsters Made of?
A Scientific Exploration through Performance

Characters

Wendy	Bob
Einstein	Prop Person
Jester 1 (J1)	Shadow Monster
Jester 2 (J2)	Keyboard Player

Note: The Prop Person and Keyboard Player may be visible or invisible on stage during the play depending on the Director.

Setting

Wendy's bedroom. On stage right is a bed crowded with pillows, blanket, bed skirt (to conceal Einstein under the bed), stuffed animals, and science textbooks. A copy of *Scientific American*, *Goosebumps*, and a flashlight. A large pink bunny rabbit sits in the middle of the bed. A portable stereo is on the floor. A pair of jeans and two sweatshirts. On stage left is a closet door. Upstage is a large white screen with a chair hidden behind it. Far stage left is a keyboard and chair.

Lights: *Full stage*
Sound: *Music*
Lights: *Lights dim; blackout*
Sound: *Music fade out*

Jester Introduction

Jesters, on opposite sides of the stage, poke their heads out from behind the curtain from opposite sides of the stage. Retreat. Poke out a leg. Retreat. Then poke out their heads and make funny faces at each other. Retreat. Then dash across the stage with stride jumps meeting at centerstage. Disappear. Jester 1 leaps out on stage.

J1. Lights! (*Full stage.*)
J2. Sound! (*Claps; J1 joins J2; they dance across stage, bump into each other, and fall down.*)

J1. Is that what monsters are made of? (*J2 shrugs; Jesters stand up, brush each other off, straighten hats, and come downstage center.*)
    Lights: *Side spotlights on Jesters; stage lights off*
J1. We'd like to welcome you one and all. We hope you really have a ball.
J2. Everyone here has a special part. You've all been chosen because you're so smart.
J1. So let's get you into the groove. And we'll teach you the moves.
J2. We'll snap you into form. Your sound effects will create a storm.
J1. But before we get you all riled up …
J2. We need a way to shut you up!!!
J1. How rude! There are a couple of signals you'll need to know. When to stop and …
J2. When to go!
J1. So if you're making sound and we do this signal (*Raises arms; audience calls out: Louder!*). That's right, you get louder. And if we do this signal? (*Lowers arms; audience calls out: Quieter!*). That's right, you get, softer. And if we do this signal? (*Arms crossed in front and then out to sides, umpire style*).
J2. Safe at first!!
J1. This isn't a baseball game! What does it really mean? (*J2 & audience yell: Stop!*)
J2. OK, let's get going! We'll get that wind blowing.
J1. When I walk in front of you, do what I do!
Jesters lead audience in making a windstorm. J1 leads audience from right to left, rubbing hands together. J2 waits and then follows tapping 4 fingers (2 on each hand) together. J1 returns to starting position and has audience tapping 8 fingers together. J2 returns to starting position

1

2

and has audience clapping thighs louder. J2 returns in reverse actions to starting position and begins blowing, making wind sounds. J1 stops clapping thighs and joins in wind sounds with audience. J2 makes the stop signal.

J1. I think you're ready to go.
J2. I think it's time to start the show!!!

Lights: Blackout

Scene 1: Bedroom

Lights: Full stage
Sound: Loud music
Setting: Wendy's bedroom; Wendy enters dancing

J1. Wendy, turn down that music!
J2. You should be doing your science homework.

WENDY, sits on bed and picks up stuffed rabbit. You know, Einstein, I really hate this science homework. All we do is read from big fat textbooks and memorize formulas and definitions like angle of refraction and angle of reflection. It's all scientific bafflegab to me! Why don't they just use normal language? Science is so boring! (Wendy turns music off.)

Jesters offstage or far stage right and left

J1. Wendy! It's time for bed.
J2. Clean up your room.
J1. Pick up your clothes.
J2. Brush your teeth.
J1. Wash your face.
J2. No reading tonight.
J1. Turn off your lights.

WENDY. Okay! Okay! (Quickly tidies up her bed, accidentally tosses bunny under bed along with books and other stuffed animals.) Books and animals under the bed. Pick up the clothes. Clothes on the chair. One! (Throws sweatshirt.) Two! (Throws jeans.) Three! (Throws sweatshirt.)

Lights: Backlight on screen. Dim front lights on stage. See

shadow of chair on screen. Audience sees shadow of clothes land on shadow chair thrown by Prop Person unseen offstage.

Backlight off. Stage lights come up

WENDY. Einstein where are you? Come on, it's time for bed! Where are you hiding now?

EINSTEIN, from under the bed. Achoooo!! (Climbs out from under the bed, sneezes again, dusts off fur.)

WENDY. What are you doing hiding under the bed, Einstein?

EINSTEIN. You threw me under the bed! I keep telling you not to do that! It's really dusty under there! I feel like a dust bunny!

WENDY. Sorry, Einstein. I don't know why there is so much dust under the bed all the time.

EINSTEIN. Well, Wendy, there is a scientific explanation for that.

WENDY. There is? Let's hear it!

EINSTEIN. Dust is actually made from tiny particles of dead skin.

WENDY. You mean, our skin dies and flakes off to become dust floating around?

EINSTEIN. Precisely!

WENDY. Yuck!

EINSTEIN. And what makes you sneeze is the dust mites that feed and live on the tiny particles of dust.

WENDY. You mean, there are tiny little bugs that live on our dust and they are the things that make us sneeze?

EINSTEIN. Exactly!

WENDY. EWW! Yuck! Let's just go to bed, OK?

EINSTEIN. OK. (Wendy & Einstein climb into bed under the blanket; Wendy turns on flashlight.)

Lights: Spotlight on bed

WENDY. So, what are you reading anyways?

EINSTEIN. Scientific American, latest issue. What are you reading?

WENDY, holds flashlight under her face. Goosebumps!

EINSTEIN. That's too scary for me!
WENDY. You're a chicken.
EINSTEIN. I'm not a chicken. I'm a bunny!
J1. Wendy, I know you are reading in there!
J2. Turn that flashlight out!
J1. Good night.
J2. Sweet dreams.
J1 & J2. Love you.
WENDY. How did Mom know I was reading? Moms have extra eyes. Come on, Einstein. Lights out.
Lights:	Spotlight on bed dims
WENDY. Good night, Einstein!
EINSTEIN. Good night, Wendy. Sleep tight. Pleasant dreams.
WENDY. Hey, Einstein, do bunnies dream in colour?
EINSTEIN. Snore!!
Lights:	Slides on screen. Slide #1: Carrots. Slides may be shown by Prop Person from slide projector, visible to audience
EINSTEIN. I love carrots!
Lights:	Slide #2: Einstein image
EINSTEIN. Albert Einstein! My hero! $E = mc^2$
Lights:	Slide #3: Jonathan Taylor Thomas (or any famous pop or movie star)
WENDY. Jonathan Taylor Thomas! He's so cute!
Lights:	Slide #4: Coyote! (Nightmare slide)
EINSTEIN. Wyle. E. Coyote! Help!
Lights:	Spotlight on bed
WENDY. Einstein, wake up! You're having a nightmare!
EINSTEIN. It was horrible! Just horrible!
WENDY. You're okay now. Try to go back to sleep, OK?
EINSTEIN. OK. (Wendy & Einstein settle down to go to sleep again.)
Lights:	Spotlight on bed dims

5

Sound:	Shadow Monster theme song
WENDY. That sounded weird. I wonder what it is.
Lights:	Backlight on screen; sees shadow of chair and pile of clothes
Shadow of monster sitting on chair slowly stands up
WENDY. That sounded weird. I wonder what it is.
Lights:	Backlight on screen; sees shadow fill the screen; Wendy shakes Einstein
Sound:	Monster roar
EINSTEIN. What is it? What's wrong?
WENDY. Don't look! I think there's something behind us! (Einstein turns slowly and sees monster. Monster roars. Both scream and dive under blanket.)
Lights:	Backlight off, monster disappears; spotlight on bed
Wendy & Einstein peek out from under sheets.
WENDY. Is it gone? Do you see anything?
EINSTEIN. No.
WENDY. Do you hear anything?
EINSTEIN. No.
WENDY. I think it's gone.
EINSTEIN. I sure hope so! I hate monsters!
WENDY. I think we should look for it!
EINSTEIN. Oh, do you really think that's a good idea?
WENDY. I won't be able to sleep if we don't find it! Come on! (Drags Einstein out of bed and turns on flashlight.)
Lights:	Stage lights on dim
WENDY. Now, where do monsters like to hide?
EINSTEIN. Under the bed?
WENDY. That's right! Let's check there first!
EINSTEIN. OK, you check.
WENDY. No, you check.
EINSTEIN. No, you check!
WENDY. Rock, paper, scissors?

6

Wendy & Einstein do rock—paper—scissors (a hand game). Players close fists; count to three then revealp either a fist (rock), a flat hand (paper), or two fingers (scissors). Rock beats scissors; paper beats rock; scissors beat paper. After three tries, Einstein wins.

EINSTEIN. You lose! You check under the bed!

WENDY. OK, OK.

EINSTEIN. Do you see anything?

WENDY. No, just THIS! (*Scares Einstein with a stuffed toy.*)

EINSTEIN. WAHHH! (*Jumps back.*)

WENDY. Ha! Gotcha! No monster though.

EINSTEIN. Great! No monster! Let's go back to bed.

WENDY. But it must be somewhere! We have to keep looking! Where else do monsters like to hide?

EINSTEIN. In the closet?

WENDY. Great idea! You check the closet!

EINSTEIN. No, you check.

WENDY. No, you check this time!

EINSTEIN. No, no. You check!

WENDY. Rock, paper, scissors?

EINSTEIN. OK! (*Wendy & Einstein play rock—paper—scissors three times, and Einstein wins.*)

EINSTEIN. You lose! You check the closet!

WENDY. OK, but come right behind me! (*Opens door and enters the closet with flashlight. Einstein glances back over his shoulder into bedroom and then follows Wendy.*)

Lights: Blackout

Prop Person removes the bed, stereo, and all stuffed animals and books. Offstage, Prop Person puts broccoli under the pillow.

7

Scene 2: Land of Light

Sound: Land of Light signature sound; continues through blackout into next scene

Lights: Three colours — red, blue, and green — on screen

Wendy & Einstein enter Land of Light. Wendy is still holding her flashlight. Einstein holds paws over eyes

WENDY. Where are we?

EINSTEIN. I don't know.

WENDY. Do you see anything?

EINSTEIN. Nope.

WENDY. Einstein! You are supposed to look! How can you see anything like that? Now, let's look for that monster, OK?

EINSTEIN. OK (*Einstein & Wendy look around; Einstein discovers coloured shadows of himself on the screen.*) Hey, Wendy! Come and look at this!

WENDY. Hey, this is neat! Coloured shadows!

EINSTEIN. See here, my shadow is blue and yellow and pink!

WENDY. And mine is red and green and orange! Why are there coloured shadows, Einstein?

EINSTEIN. There is a scientific explanation for that, Wendy!

WENDY. There is?

EINSTEIN. Yes. You see, coloured shadows are a perception of distinguishing varying ways in which homogenous patches of light such as hue, saturation, and lightness appear to a subject.

WENDY. So you mean if you have a red light, a green light, and a blue light shining together on something, you can make coloured shadows?

EINSTEIN. Umm, I think so. (*Scratches his ear.*)

WENDY. Cool! This is really neat, but we better look for that monster. (*Walks stage right, with flashlight looking for monster.*)

WENDY. Einstein, are you looking?

8

EINSTEIN. Yup! (Continues shadow dancing.)

WENDY. I said to stop fooling around!

EINSTEIN. Okay!

Lights: Coloured lights off, stage lights dim, a low front spotlight is turned on

Einstein turns around and faces audience; he walks toward the light; his shadow gets bigger; he looks back over his shoulder and sees shadow.

EINSTEIN. AHHHHH! There's a giant bunny after me! Don't panic! Stay calm! There must be a scientific explanation for this? ... Ah, yes! When an object intercepts the light falling on a surface, then the size of the shadow will be dependent upon the distance that the object is away from the light source! Whew! (*Turns around and sees his giant shadow.*) It's still there! Wendy! Come over here! Quick!

WENDY. What's wrong, Einstein? Calm down!

EINSTEIN. There is a giant bunny right behind me and he is going to get me!!

WENDY. Einstein, that's not a giant bunny!

EINSTEIN. No?

WENDY. No! It's just this light making your shadow big!

EINSTEIN. Really?

WENDY. Yes! Now, just back up, OK?

EINSTEIN. You sure about this?

WENDY. Just go, Einstein! Nothing is going to get you! (*Einstein slowly backs away from the light, toward the screen; his shadow gets smaller.*) Now, turn around.

EINSTEIN. It is small! It's just me!

WENDY. Yeah! Told you it was just you and the light!

EINSTEIN. That's just what I said!

WENDY. Now that's figured out, let's go and find that monster, okay? (*Wendy & Einstein freeze.*)

Lights: Blackout; Prop Person takes closet door offstage

9

Jester Interlude: Shadow Tag

Jesters play shadow tag with flashlights in auditorium, running up and down the aisles.

J2. Flashlight tag. You're it! Hey where did you go? How did you know?

J1. Achooo!

J2. There you are! You're it!

J1. I'm going to get you.

J2, shines light on an audience member. Boo!

J1, catches J2. You're it!!

J2, finds J1 sitting in chair. Aha! Heh, you're supposed to be working you know.

J1. OK, meet you at the front!

Lights: Side spotlights on Jesters

Jesters play with lights and their shadows to illustrate

J2. If the shadow is here, where is the light? Back, front, left, or right? Big and small, short and tall. We see them on the wall. On the ground, all around, shadows can be found.

J1. When the light is bright, the shadow is dark; but it's not always black, that's a fact.

Lights: Blackout; Jesters use flashlights or use coloured spotlights

J2. With blue light, you're quite the sight. Try it with red, it'll knock you dead. Green is keen.

Lights: Side spotlights on

J1. But whether it's big or small, remember one and all ...

Lights: Side spotlights off; Jesters shine flashlights at the audience

J2. The message this sends, is IT DEPENDS!!!

Lights: Jesters shine flashlights in circles; blackout; stage lights on

Wendy & Einstein on stage.

WENDY. Are you still looking for that monster?

10

EINSTEIN. Yup! (Dances with his shadow on screen.)

WENDY. No fooling around now!

Lights: Backlight on screen. Create shadow of Keyboard Player that should be same size and position as Einstein's shadow. Prop Person stands on chair and holds spot at waist level. Front lights dim.

EINSTEIN. Left arm up, left shadow goes up. Right arm up, right shadow up. Left arm up, right shadow goes up? That's weird. Right arm up, left shadow goes up? That's not right! Wendy, come over here!

WENDY. What is it, Einstein?

EINSTEIN. Look at this shadow! Left arm up, right shadow goes up.

WENDY. That's not right!

EINSTEIN. Right arm up, left shadow goes up!

WENDY. That is really weird! See what happens when you go down.

Einstein crouches down and shadow changes into Shadow Monster. To do this, Prop Person gets off chair and holds spotlight at feet of Keyboard Player.

Sound: Monster roar

EINSTEIN. Run, Wendy! It's that monster! Run! Run for your life!

WENDY. Follow me! This way, Einstein!

Chase Scene. Wendy & Einstein run back and forth across stage followed by monster's shadow. Lights: Strobe light is turned on; backlight is off. Wendy & Einstein run in slow motion. Sound: Keyboard Player sits at keyboard; note played at the same frequency as light strobe. Wendy & Einstein collide and fall down. Lights: Strobe light is turned off; blackout. Sound: Repeating note, crescendo to musical flourish for Land of Sound.

Lights: Full stage lights slowly come up cued with sound for Land of Sound

11

Scene 3: Land of Sound

Prop Person backstage sets up red spot to flash on and off during explosion at end of scene.

WENDY. Einstein, are you OK?

EINSTEIN. I think so. Are you OK?

WENDY. Yes. Where are we?

Einstein. I don't know.

WENDY. Do you hear anything?

WENDY. Einstein, you can't hear anything if you cover your ears. Now do you hear anybody?

EINSTEIN. I don't hear anybody, body, body …

WENDY. Hey, there's an echo! Hello …

J1, from back of auditorium. Hello.

J2. Hello.

WENDY. Cool! Try it out, Einstein!

Prop Person backstage EINSTEIN. Hello.

J1. Hello.

J2. Hello.

WENDY. Do another one, Einstein.

EINSTEIN. You're Awesome!

J1. You're Awesome!

J2. You're Awesome!

EINSTEIN. You're great!

J1. You're great!

J2. You're great! You're wonderful! (Jesters begin echo complimenting each other.)

J1. Yes, I am! You're fantastic!

J2. Yes, I am! You're fantastic!

EINSTEIN. Wendy, there's a scientific explanation for this.

WENDY. There is?

EINSTEIN. Yup! You see, sound is the motion of oscillating particles moving parallel through a medium in the direction of the propagation of its horizontal sound wave. Therefore, an

12

echo is a reflection of a horizontal sound wave by a surface so that a weaker version of this is detected shortly after the original sound wave!

WENDY. Einstein, eat a carrot. (*Hands Einstein a carrot; he bites it 3 times and with each bite is a different sound.*)

    Sound: Carrot crunching noise.

EINSTEIN. Mmmmm. Good carrot.

    Sound: Car horn.

EINSTEIN. That didn't sound right!

WENDY. Try it again. (*Einstein & Wendy ad lib.*)

    Sound: Loud ticking.

EINSTEIN. Oh no! Wendy, it's a bomb! Here you take it!

WENDY. No, you take it!

EINSTEIN. You take it!

WENDY. Look out! It's going to explode! (*Throws carrot offstage. Einstein & Wendy dive for cover, putting hands over ears.*)

    Sound: Loud explosion

    Lights: Flash of white and red lights for explosion;

    Prop Person flickers red spot on and off

    Wendy & Einstein blasted offstage.

    Lights: Blackout

Jesters Interlude: The Game Show: Guess that Sound!

    Lights: House lights

    Sound: Game show music

J2. It's that time for Guess that Sound!

J1. All we need are a few volunteers; all you need are 2 ears!

J2. But first, our lovely assistant Bob will help show the prizes.

    Sound: Bob's entrance music

    Bob comes on stage wearing an evening gown and hiking boots with big socks; his job is to model the prizes.

J2. Our first prize: contestants will win an all-expense paid trip to Disneyland. Say hi to Mickey and Minnie while you're there!

13

Prop Person hands enormous bag of Puppy Chow to Bob.

J1. Runners-up will receive a year's supply of everyone's favourite: Puppy Chow! Yum-yum, Bob!

J2. Listen to the sound and mull it around. Guess what it's about. Feel free to shout it out!

    Audience hears a sound, and a group guesses what it is (e.g., doorbell, train whistle, baby crying, etc.). J1 & J2 mime the actions that make each sound. J1 identifies Group 2.

J1. OK. Group 2, you know what to do. It could be a train, a whistle, or drums; either way, here it comes.

J2. That's right! It's a ....

    Audience hears a sound and Group 2 guesses what it is. J2 identifies Group 3.

J2. OK, Group 3, there's not much to do as you can see. Listen really well. I know you'll be able to tell.

    Audience hears a sound and Group 3 guesses what it is.

J2. You guessed it! It's a .... Way to go, Group 3!

J1. Drum roll, please. (*Audience claps hands on thighs.*) Thank you for playing Guess that Sound. And now for the prizes!!

J2. Bob, what do you have for our lucky contestants? Come on out here, Bob. Bob? (*Looks for Bob behind the curtains. Peers backstage. Checks chairs in the audience. Returns to J1 and whispers agitatedly in J1's ear.*)

J1. It would appear that both Bob and the tickets for Disneyland are missing!

J2. A big round of applause for all the contestants! TV! What else can we say?

J1. I know, back to the play!

    Einstein & Wendy stagger back on stage.

WENDY. Einstein, are you alright?

EINSTEIN. Yea, but what was that?

WENDY. That was an exploding carrot bomb!

14

EINSTEIN. I'm never going to eat carrots again. They are potentially dangerous!

WENDY. But Einstein, what will you eat then?

EINSTEIN. Hmmm.... I know! Broccoli!

WENDY. Great idea! Now let's look for that monster.

EINSTEIN. You do that, I'm getting out of here. (Tries to walk offstage. Use stage steps if available.)

    Sound: Keyboard Player makes sounds according to Einstein's footsteps, one per step

EINSTEIN. Yikes! (Jumps back and then decides to test the step carefully. Gets braver and walks up and down the steps or across the stage. Each footstep has a corresponding note on the keyboard. He gets the hang of it and finds "dancing music with his feet" fun.) Hey! Look, Wendy! (Dances on steps.) This is really fun! Try it out!

WENDY. OK. Here goes! (Joins in. Keyboard Player plays as Wendy & Einstein dance with the music.)

WENDY. Gee, Einstein. I'm getting tired.

EINSTEIN. Me too!

WENDY. It's time for the grand finale.

EINSTEIN. Let's do it! (Wendy & Einstein dance wildly and then jump. Instead of a chord, they hear the monster roar.) Ahhh! It's the monster!

WENDY. This way, Einstein! (They run off stage.)

    Lights: Stage lights dim; side spotlights on

Jester's Interlude: The Windstorm

J1. We've got to do something with that monster!

J2. I know! The storm, Let's blow the monster away!

J1 & J2. The storm, the storm!!! (Jesters run frantically around and then realize they can use the audience.)

J1. OK, man your stations. This is not a drill!

J2. This is an emergency. This is what you've trained for!

    Jesters do windstorm with audience. (See windstorm

15

instruction to audience in beginning of play).

    Prop Person carries fans with ribbons on stage and sets them up. When Jesters start doing wind with audience, Prop Person turns on fans and Shadow Monster is blown offstage.

    Sound: Sound of wind added to audience's blowing

    Jesters signal audience to stop blowing and sound effect of blowing takes over.

Scene 4: Land of Movement

    Lights: Side spotlights off; stage lights full up

    Wendy & Einstein return to stage with hair and ears messed up. Wind is blowing hard as they try to walk toward fans.

WENDY. What's with the wind?

EINSTEIN. I think it's a windstorm.

WENDY. A windstorm? Einstein, why is it so windy?

EINSTEIN. Actually, there's a scientific explanation for that. Wind is air in motion, caused by the rotating surface on the Earth.

WENDY. What? I can't hear you.

EINSTEIN. It presents a 3-dimensional structure, with its horizontal component being considerably greater than its vertical components (spins in a circle). And the wind that is produced is an attempt by the Earth's atmosphere to maintain equilibrium.

WENDY. Einstein, that is interesting, but this wind's blowing me away! (They get blown away from fans and get stuck on stage left wall or curtain.)

    Sound: Wind noise stops; Prop Person removes fans from stage

WENDY. I'm stuck!

EINSTEIN. Me too! Can you get unstuck?

WENDY. I think so. (They take a few steps, making slurping sounds.) It sure is hard walking through this stuff! What do you

16

think this is?

EINSTEIN. Some kind of mud or glue!

WENDY. I'm tired! I just want to go home!

EINSTEIN. Me too!

WENDY. Which way should we go?

Prop Person holds out fishing rod with broccoli on end of line on stage right.

EINSTEIN. To the broccoli!

WENDY. Broccoli? OK, you first!

EINSTEIN. Follow me! (They continue walking, making slurping sounds.)

Sound: Running water

EINSTEIN. Hey, do you hear something?

WENDY. Sounds like water, or a river.

EINSTEIN. What is a river doing here?

WENDY. How are we going to cross it?

EINSTEIN. I don't know! I can't swim. I'm a bunny! Wait a minute! There's a log! Let's use that to get across.

WENDY. OK. You go first. Be careful! I'm right behind you! (Jumps on the log behind Einstein. Log bounces up and down, in mime; they try not to lose their balance.)

WENDY. Ooops! Einstein, watch your balance on the log. It's a little shaky!

EINSTEIN. OK. Wendy, remember to keep your centre of gravity low and base of support wide! I think can reach the broccoli. I've almost got it!

WENDY. Be careful, Einstein.

EINSTEIN. I'm losing my balance! I'm falling! (Tries to reach for the broccoli and falls into the water.)

Sound: Splash

Lights: Blue lights; bubbles; reflected light waves on the screen

17

EINSTEIN. I'm drowning! Help me! Somebody help me!

WENDY. Hang on, Einstein! I'm coming in! (Holds nose and jumps in; sound of splash. Wendy swims [breaststroke] toward Einstein.)

Sound: Jaws music

Jester 2 holding shark fin swims across the front of stage in front of Einstein & Wendy.

EINSTEIN. Oh no! It's a shark! I'm shark bait. Swim Wendy, save yourself! (Keeps trying to swim, like climbing a ladder. Wendy swims front crawl to escape.)

Lights: Blue lights dim; side stage spots

Jesters Interlude: Saving the Drowning Bunny

Jester 1 & Jester 2 are playing rock—paper—scissors. J1 sees Einstein drowning. J2 keeps playing the game without J1.

J1. Stop fooling around! They have enough problems without you! How can we help Einstein? We need something that floats!

J2, grabs bag of Puppy Chow from Prop Person. How about this? Will it float?

J2 shows audience different items to rescue Wendy & Einstein (Puppy Chow bag, rubber ducky, large rock, etc.) Meanwhile, J1 asks audience with "thumbs up or down" if items will work. Prop Person could be on stage handing items taken from large prop bag to J2.

J2, points to a child in audience. How about you? Do you float?

J2. Bob would know what to do!

Sound: Bob's theme song

Bob dressed in shorts and bright shirt enters from the back of the audience, rolling a pool toy tire tube.

Lights: Spotlight follows Bob to front and centre stage

J1. Nice tan, Bob!

J2. How was Disneyland, Bob?

18

J1. Those tickets were for them, Bob! (Points to audience.)

J2. Way to go, Bob!

J1 & J2, to audience. Help us count! One, two, three!

Lights: Blue stage lights up; side spots off

EINSTEIN. Help! Wendy!

Prop Person tosses Bob's pool toy tire tube to Wendy from off stage.

WENDY, catches the tire. Einstein! Grab on to the tire! Hold on, are you OK?

EINSTEIN. Whew! Now what are we going to do, Wendy?

WENDY. I don't know. We've got to get out of here somehow.

Lights: Blue lights dim. Side spots on Jesters with flashlights.

J1. They're going to be floating around in the water forever. They'll be like shriveled raisins.

J2. Mmmmm, raisins. I like raisins.

J1. What can we do? (Jesters shine flashlights around auditorium and see plug on stage wall right.) Hey, there's a plug; go pull it out.

Lights: Side spotlight follows J2 to plug on auditorium wall right.

J2. The plug! I've got it!

J1. One, Two, Three! Pull!!! (J2 collapses on the floor; plug still in the wall.)

J2. Help me pull out the plug! (J2 & J1 grab onto plug together, struggle.)

J1. Pull harder! I know! Let's all count together (with audience) 1...2...3...

J1 & J2. PULL!

J1 & 2 pull out the plug and water swirls down the drain. Wendy & Einstein twirl around in the water faster and faster. Sound of water going down the drain. Wendy & Einstein collapse on the floor.

Lights: Blackout; Prop Person places closet door stage

19

left; lights dim

Wendy & Einstein pick themselves off floor and wring themselves out and shake off water.

EINSTEIN. I'm all wet. (Wrings out ears, shakes bunny tail.)

WENDY. Me too! Hey! Stop shaking water on me!

EINSTEIN. Ooops! Sorry.

WENDY. I think it's time to get back home.

EINSTEIN. Sure. I agree.

WENDY. But how do we get back home?

EINSTEIN. Hey, isn't that our closet door?

WENDY. I think it is! Let's go and check it out!

EINSTEIN. OK, you check!

WENDY. No, you check!

EINSTEIN. No, no! You check!

WENDY. Rock, paper, scissors?

EINSTEIN. OK.

WENDY. I've got to win this time!

Wendy & Einstein play rock—paper—scissors. Einstein wins the 3rd time with "giant squid kills all." (Hand up, fingers down, shaped like a squid, wriggle fingers). They go through the door.

Sound: Creaking door followed by musical flourish for Land of Light into next scene

Lights: Blackout

Scene 5: Back to Bedroom

Wendy's bed on stage. A suitcase is hidden under the bed. Stereo on floor. Wendy & Einstein enter through closet door, Wendy leading with a flashlight.

Lights: Flashing lights; stage lights on dim

WENDY. Where are we?

EINSTEIN. I don't know

WENDY. Do you see anything? Is the monster here?

EINSTEIN. I don't see anything (Paw covers eyes.)

20

WENDY. You are supposed to look, Einstein! Now where can we be?

EINSTEIN. Isn't that your bed, Wendy?

WENDY. I think it is! And this is my stuff! We're home!

EINSTEIN. We're home!

Lights: Stage lights full

Wendy & Einstein celebrate with cheering and high-fives.

WENDY. What?

EINSTEIN. Is that monster here?

WENDY. I don't know. I think we better check. I'll check under the bed; you check over there. (*Looks under bed; Einstein looks under his legs.*) Do you see anything?

EINSTEIN. Nothing over here. Do you see anything?

Sound: Monster roar

EINSTEIN. You left the closet door open!

WENDY. You were supposed to close the door! You came in last!

EINSTEIN. No, you were!

WENDY. Well, what are we going to do now? Oh, why is there a monster in my room?

EINSTEIN. Well, Wendy, there's a scientific explanation for that.

WENDY. There is?

EINSTEIN. Yup

Lights: Blackout. Theory slide #1

EINSTEIN. You see. Monsters = L + S + M.

Lights: Theory slide #2

EINSTEIN. Therefore, monsters equal light plus sound plus movement.

WENDY. Which means?

Sound: Monster roar

Lights: Stage lights up

EINSTEIN. Which means this bunny's outa here! (*Goes over to bed and pulls out his suitcase.*)

21

WENDY. But where are you going?

EINSTEIN. I'm heading down that bunny trail!

WENDY. Hold on a minute! You don't have to go!

EINSTEIN. I don't?

WENDY. Look, we have a theory, right?

EINSTEIN. We do?

WENDY. Yes! And if we know what monsters are made of, we can make our own monster.

EINSTEIN. We can?

WENDY. Sure! And we can scare that monster away!

EINSTEIN. You sure?

WENDY. Well, we can try. What do you need first? Light! Let me see ... I know! I can use my flashlight! What do we need next? Sound! Let's try some of my music!

Sound: Song #1 Baby Beluga by Raffi

Wendy & Einstein start dancing.

WENDY. I don't think this will work.

EINSTEIN. Great song though!

Sound: Song #2 Einstein's shower song

WENDY. Einstein! You've been messing with my music!

EINSTEIN. Singing in the rain. I'm singing in the rain.

WENDY. This next one should work.

Sound: Song #3 Einstein's combat song (upbeat music)

WENDY. Great! This will work!

EINSTEIN. Sounds good!

WENDY. So, what else do we need? Movement!

EINSTEIN. What can we do?

WENDY. I know! You've got big ears, move them around, and you've got big arms, so move them around! You've got big feet! Move them around. (*Shows Einstein how to move.*)

EINSTEIN. We're moving around!

WENDY. Yea! You are doing great!

EINSTEIN. We can get that monster!

22

WENDY. You can get that monster!

EINSTEIN. Yea!

WENDY. Go, Einstein! You can get that monster!

EINSTEIN. Yea! Bring him on!

WENDY. Go, Einstein!

EINSTEIN, *starts to leave the stage but turns back.* You sure about this?

WENDY. Of course, I'm sure! You are great, awesome! If you don't, I'll … I'll never give you broccoli!

EINSTEIN. For broccoli, I'll do anything! (*Leaves stage.*)

WENDY. Are you ready Einstein?

EINSTEIN. Yup!

WENDY. OK! First, I need light! I'll put my flashlight right here.

    Lights: *Stage lights dim to back light on shadow of monster on screen*

WENDY. And I'm going to turn on the music!

    Sound: *Song #3 Einstein's combat song; monster roar*

WENDY. Now, Einstein start moving around!

    Shadow of monster on stage left. Einstein enters stage right. Shadow of monster is much bigger than Einstein's shadow. Einstein tries to punch Shadow Monster who just put his hand on Einstein's head so punches can't reach him.

WENDY. Uh-oh. It isn't working! I have to move my flashlight closer! And turn on music louder! Keep moving, Einstein!

    Shadow Monster moves away from back spot toward back screen and becomes smaller. Einstein moves toward back spot and becomes larger.

J1 & J2. Einstein! Einstein!

    Shadow Monster sees the size of Einstein's shadow and runs away.

    Sound: *Victory song music*

    Wendy & Einstein dance their victory dance. Einstein still behind screen, then comes back into bedroom from behind screen.

    Lights: *Full stage lights; back light off*

WENDY. You were awesome!

EINSTEIN. We scared the monster!

WENDY. Yahoo!

J1. Wendy! Turn down that music!

J2. You should be sleeping!

WENDY. I don't believe it! If Mom only knew what we've been through!

EINSTEIN. Coloured shadows!

WENDY. Exploding carrots!

EINSTEIN. Sticky stuff!

WENDY. Windstorms!

EINSTEIN. Drowning and sharks!

WENDY & EINSTEIN. And monsters!

WENDY. Whew! Am I tired!

EINSTEIN. Me too!

WENDY. Let's go back to bed, OK?

EINSTEIN. OK!

    Wendy & Einstein climb into bed.

WENDY. Good night, Einstein.

EINSTEIN. Good night, Wendy.

    Lights: *Stage lights dim*

    Sound: *Monster theme music*

    Wendy & Einstein sit up in bed.

SHADOW MONSTER, *speaks from offstage.* Light, sound, moving around. That's a good theory. But that's not all monsters are made of!

    Wendy & Einstein hide under covers, shaking.

    Lights: *Blackout*

    Sound: *Monster roar*

    Curtain close

4.4 Exploring Science through Drama: Interdisciplinary Learning and Creative Action

As we watch the play again—now documented digitally and flickering on our computer monitor, a shadow momentarily falls across the screen. What monsters are we still fighting in science education? This question encourages all of us to identify and face possible monsters in science education. How might we create a welcoming space of possibility to invite diverse beginning teachers into the exploration of science and science teaching? How do we attend to the reluctant learner or the learner for whom science is a shadow monster?

On reflection these many years later, we recognize that (a) the scholarship that arose from our interdisciplinary collaboration and (b) the educational concepts that guided our teaching of science education through performance-and-story resonate in today's educational landscape. These educational practices reflect cross-curricular learning through drama (Fels & Belliveau, 2008; Tarlington & Verriour, 1991). Specifically, the integration of drama and science spans decades (Çokadar & Yilmaz, 2010; Meyer & Fels, 1998; Najami et al., 2019; Ødegaard, 2003; Palmer, 1999; Raphael & White, 2021; White et al., 2021).

British Columbia provincial curriculum documents encourage inquiry through a diversity of teaching strategies and multiple literacies, focusing on the "Big Idea" that recognizes interconnections between lived experience, different modes of inquiry, and creative and collaborative engagement (BC Ministry of Education, 2022a, 2022b, 2022c). Core competencies of communication, thinking, personal and social awareness, responsibility and identify may be incorporated across multiple subject areas, including science and arts education (BC Ministry of Education, 2022b, 2022c). It is encouraging that drama and science continue to be linked as meaningful inter- or transdisciplinary ventures (Raphael & White, 2021) to achieve these goals.

Through the years, studies investigating science teaching consistently encourage the implementation of drama as a strategy to facilitate comprehension of science concepts and enhanced engagement in science learning (Abed, 2016; Henderson & King, 2021; Hendrix et al., 2012; McGregor, 2012). Drama may be incorporated in a science classroom in multiple ways that can be beneficial for student learning. Dramatization of science concepts can facilitate student understanding and enthusiasm (Saricayir, 2010); scripts, improvisations, simulations, and role play can be implemented so that students might consider the impact of science and technology on society and on their individual lives (Baskerville & Anderson, 2021; Swanson, 2021). Learning through drama is embodied, affective, relational, and collaborative, and requires decision-making, problem-solving and creative, critical, and reflective thinking—all qualities one wishes to develop in tomorrow's scientists.

Our relationship with each other and our students, and the children with their teachers, was a relationship and praxis of what Freire (1970/1995) called *horizontality,* which recognizes that knowing and learning dwells within community and in shared interaction with each other—a liberatory praxis that recognizes communal responsibility for shared learning. We embodied an emergent curriculum (Bava et al.,

2022; Davis et al., 2000) that consistently surprised, challenged, and delighted. Interdisciplinary collaboration (Kaufman & Brooks, 1996; Larraz Rada et al., 2014) and creativity through improvisation (Duffy, 2006; Egan et al., 2015; Nachmanovitch, 1990) were at the heart of our project; we engaged as co-teachers, co-scientists, and co-performers in learning and interaction with our students. Creativity or creative action embodies imagination and action in educational spaces of play and inquiry (Fels, 2022; Hatt, 2018). Hatt (2018) stated, "imagination ignites curiosity and inventiveness and activates the powerful process of creativity that engenders images of the possible and that leads to innovative action in the life-world of 21st-century education" (p. 127).

4.5 Epilogue: Defeating the Monster

Ours was a philosophical stance that dwelled in hermeneutic phenomenology. The essences of being— phenomena encountered and experienced, perceived and interpreted—call us to awareness of who we are in co-evolving interactions within and between human and non-human; as educators and researchers, we attend to the ethics of what is, what matters, that is our living experience (Jardin, 2021; Russon, 1994). As science educators, performers, learners, we explored light, sound, and movement and were informed by our inquiry of phenomena as lived in interaction and curiosity (Meyer, 2006; Feher & Rice-Meyer, 1992; Gallagher, 1986; van Manen, 1990). Ours, then, was a living inquiry that called us to attend thoughtfully, playfully, critically to our experiences together—skating, making shadows, improvising scripts, performing a play—through an exquisite exploration and opportunity of time, place, and being in relationship.

Our adventure together was a performative inquiry, where challenging pedagogical scripts and questioning what is possible in a science education classroom invited us to reimagine our roles as science educators, artists, and learners. Ours was a language of experiential meaning-making—not the dense scientific vocabulary of our uncomprehending Einstein character but through the play-full language of serious exploration (i.e., What if? What matters? What happens? Who cares?) that invites us to be present with our students, investigating, creating, and storying in the midst of the unfolding mysteries of life.

What is this event we celebrate? a play—an encounter between and with education students, teachers, children, science, and drama that drew on the curiosity, imagination, and creativity in the pedagogical spaces we co-created. We explored science and drama as interaction, thus "enlarging the space of the possible" (Sumara & Davis, 1997) as we performed our play for children who came to the theatre understanding how light, sound, and movement work so that they could yell advice from the audience, actively participating in the adventures of Einstein, Wendy, and the Shadow Monster. "What we do is what we know, and ours is but one of many possible worlds. It is not a mirroring of the world, but the laying down of a world." (Varela, 1987, p. 62). Something new arrives:

When we journey into science through drama and storytelling, we begin to understand that science is realized through a creative and critical exploration of perceived and imagined phenomena, a vibrant search … of action and interaction. There are many ways to imagine a universe into being, and, through the vehicle of our imagining, to voice our interstandings. (Fels, 1999, p. 91)

Twenty-six years ago, a pink-eared six-foot bunny rolled out from under Wendy's bed and introduced us to the joy of playmaking in science. You are welcome to perform our play with your students and, in so doing, experience the creativity that is science. We invite you into the play that is science education. Close your eyes for a moment and visualize our floppy-eared Einstein and Wendy in your classroom. Imagine inviting your science students to work in tandem with drama students to perform *Light, Sound, Movin' Around: What are monsters made of?* Imagine preservice education students bringing the play to children in your local elementary school.

We invite you and your students to breathe and reinterpret our play into life. Imagine calling your students to creative action and activism through playmaking to tackle monsters such as climate change, air or water pollution, artificial intelligence, transglobal viruses, urban sprawl, marine life destruction. There, in the heart of your classroom, are plays of science yet to be written and worlds yearning to come into being.

Acknowledgements *Light, Sound, Movin' Around: What are monsters made of?* (1996) was performed at the Normand Bouchard Memorial Theatre, University of British Columbia, Vancouver, BC, in the morning and afternoon of Thursday, April 25, 1996. We are thankful for the imagination and joy of our preservice students who wrote and performed in the play. We applaud all the children and their teachers who came to the theatre to help us conquer the Shadow Monster. We are grateful to live, play, and work on the ancestral and unceded territories of the Coast Salish peoples of the Səlilwətaɬ (Tsleil-Waututh), Skwxwú7mesh (Squamish), and xʷməθkwəy̓əm (Musqueam) Nations.

References

Abed, O. (2016). Drama-based science teaching and its effect on students' understanding of scientific concepts and their attitudes towards science learning. *Journal of International Education Studies, 9*(10). 163–173. https://doi.org/10.5539/ies.v9n10p163

Baskerville, D., & Anderson, D. M. (2021). Responding to climate change: Developing primary children's capability to engage with science through drama. In P. J. White, J. Raphael, & K. van Cuylenburg (Eds.). *Science and drama: Contemporary and creative approaches to teaching and learning* (pp. 93–105). https://doi.org/10.1007/978-3-030-84401-1_6

Bava, S., Engelbrecht, A., Fels, L., Gramani, D., Grindlay, M., Johnsey, S., Kalra, A., Lin, M., Martens, M. M., & Lau, A. W. (2022). Inception of emergence. *International Journal of Education & the Arts, 23*(5). https://doi.org/10.26209/ijea23n5

British Columbia Ministry of Education. (2022a). *BC's curriculum.* https://curriculum.gov.bc.ca

British Columbia Ministry of Education. (2022b). *BC's curriculum: Arts education.* https://curriculum.gov.bc.ca/curriculum/arts-education.

British Columbia Ministry of Education. (2022c). *BC's curriculum: Science.* https://curriculum.gov.bc.ca/curriculum/science

Christiansen, H., Goulet, L., Krentz, C., & Maeers, M. (Eds.). (1997). *Recreating relationships: Collaboration and educational reform*. SUNY Press.

Clark, C., Moss, P. A., Goering, S., Herter, R. J., Lamar, B., Leonard, D., Robbins, S., Russell, M., Templin, M., & Wascha, K. (1996). Collaboration as dialogue: Teachers and researchers engaged in conversation and professional development. *American Educational Research Journal, 33*(1), 193–231. https://doi.org/10.3102/00028312033001193

Çokadar, H., & Yılmaz, G. (2010). Teaching ecosystems and matter cycles with creative drama activities. *Journal of Science Educational Technology, 19*(1), 80–89. https://doi.org/10.1007/s10956-009-9181-3

Davis, B., Sumara, D. J., & Kieren, T. E. (1996). Cognition, co-emergence, curriculum. *Journal of Curriculum Studies, 28*(2), 151–169. https://doi.org/10.1080/0022027980280203

Davis, B. Sumara, D. J., & Luce-Kapler, R. (2008). *Engaging minds: Changing teaching in complex times* (2nd ed.). Routledge.

Duffy, B. (2006). *Supporting creativity and imagination in the early years* (2nd ed.). Open University Press.

Egan, K., Judson, G., & Madej, K. (2015). *Engaging imagination and developing creativity in education*. Cambridge Scholars.

Einstein, A. (1929, October 26). What life means to Einstein: An interview by George Sylvester Viereck. *The Saturday Evening Post*. [Verified on microfilm by Quote Investigator]. https://quoteinvestigator.com/2013/01/01/einstein-imagination/

Erickson, G., & Meyer, K. (1997). Performance assessment tasks in science: What are they measuring? In K. Tobin & B. Fraser (Eds.), *International handbook of science education* (pp. 845–864). Kluwer.

Feher, E., & Rice-Meyer, K. (1992). Children's conceptions of color. *Journal of Research in Science Teaching, 29*(5), 505–520. https://doi.org/10.1002/tea.3660290506

Fels, L. (1999). *In the wind clothes dance on the wind: Performative inquiry–a (re)search methodology: Possibilities and absences within a space-moment of imagining a universe*. Doctoral dissertation, University of British Columbia. https://open.library.ubc.ca/soa/cIRcle/collections/ubctheses/831/items/1.0078144

Fels, L. (2010). Coming into presence: The unfolding of a moment. *Journal of Educational Controversy, 5*(1), Article 8. https://cedar.wwu.edu/jec/vol5/iss1/8/

Fels, L. (2012). Collecting data through performative inquiry: A tug on the sleeve. *Youth Theatre Journal, 26*(1), 50–60. https://doi.org/10.1080/08929092.2012.678209

Fels, L. (2022). Melting ICE: Igniting imagination in empty space (A performance). In B. E. Hatt (Ed.), *Crushing ICE: Short on theory, long on practical approaches to imagination creativity education* (pp. TBD). FriesenPress.

Fels, L., & Belliveau, G. (2008). *Exploring curriculum: Performative inquiry, role drama, and learning*. Pacific Educational Press/UBC Press.

Fels, L., & Meyer, K. (1997). On the edge of chaos: Co-evolving world(s) of drama and science. *Journal of Teaching Education, 9*(1), 75–81. https://doi.org/10.1080/1047621970090113

Freire, P. (1970/1995). *Pedagogy of the oppressed* (M. B. Ramos, Trans.). Continuum.

Gallagher, S. (1986). Lived body and environment. *Research in Phenomenology, 16*(1), 139–170. https://doi.org/10.1163/156916486X00103

Goulet, L., Kentz, C., & Christiansen, H. (2003). Collaboration in education: The phenomenon and process of working together. *AJER, 49*(4), 325–340. https://doi.org/10.11575/ajer.v49i4.55027

Hatt, B. E. (2018). The new ICE-age: Frozen and thawing perceptions of imagination. *Canadian Journal of Education / Revue Canadienne de l'éducation, 41*(1), 124–147. https://journals.sfu.ca/cje/index.php/cje-rce/article/view/2496

Henderson, S., & King, D. (2021). "This is the funniest lesson": The production of positive emotions during role-play in the middle years science classroom. In P. J. White, J. Raphael, & K. van Cuylenburg (Eds.). *Science and drama: Contemporary and creative approaches to teaching and learning* (pp. 179–196). Springer. https://doi.org/10.1007/978-3-030-84401-1_11

Hendrix, R., Eick, C., & Shannon, D. (2012). The integration of creative drama in an inquiry-based elementary program: The effect on student attitude and conceptual learning. *Journal of Science Teacher Education, 23*, 823–846. https://doi.org/10.1007/s10972-012-9292-1

Jardin, D. (2021). It might just be ravens writing in mid-air. *Journal of Curriculum Theorizing., 36*(1), 1–9.

Kaufman, D., & Brooks, J. (1996). Interdisciplinary collaboration in teacher education: A constructivist approach. *TESOL Quarterly, 30*(2), 231–251. https://doi.org/10.2307/3588142

Larraz Rada, V., Yez de Aldecoa, C., Gisbert Cervera, M., & Espuny Vidal, C. (2014). An interdisciplinary study in initial teacher training. *Journal of New Approaches in Educational Research, 3*(2), 67–74. https://www.learntechlib.org/p/148240/

Maturana, H. R., & Varela, F. J. (1992). *The tree of knowledge: The biological roots of human understanding* (rev. ed.). Shambhala.

McGregor, D. (2012). Dramatizing science learning: Findings from a pilot study to re-invigorate elementary science pedagogy for five-seven years olds. *International Journal of Science Education, 34*(8), 1145–1165. https://doi.org/10.1080/09500693.2012.660751

Meyer, K. (2006). Living inquiry–a gateless gate and a beach. In W. Ashton & D. Denton (Eds.), *Spirituality, ethnography, and teaching: Stories from within* (pp. 155–166). Peter Lang.

Meyer, K. (2000). Looking for science in all the wrong places. In C. E. James (Ed.), *Experiencing difference* (n.p.). Fernwood.

Meyer, K., & Fels, L. (1998). Einstein, the universe and us: Science hits the stage—performative inquiry with a co-evolving curriculum. *Journal of Curriculum Theorizing, 14*(1), 22–26.

Nachmanovitch, S. (1990). *Free play: Improvisation in life and art.* Jeremy P. Tarcher/Putnam.

Najami, N., Hugerat, M., Khalil, K., & Hofstein, A. (2019). Effectiveness of teaching science by drama. *Creative Education, 10*, 97–110. https://doi.org/10.4236/ce.2019.101007

Ødegaard, M. (2003). Dramatic science. A critical review of drama in science education. *Studies in Science Education, 39*(1), 75–101. https://doi.org/10.1080/03057260308560196

Palmer, D. H. (1999). Using dramatizations to present science concepts: Activating students' knowledge and interest in science. *Journal of College Science Teaching, 29*(3), 187–190. http://www.jstor.org/stable/42990256

Quinlan, E., & Duggleby, W. (2009). "Breaking the fourth wall": Activating hope through participatory theatre with family caregivers. *International Journal of Qualitative Studies on Health and Well-Being, 4*(4), 207–219. https://doi.org/10.3109/17482620903106660

Raphael, J., & White, P.J. (2021). Transdisciplinarity: Science and drama education developing teachers for the future. In P. J. White, J. Raphael, & K. van Cuylenburg (Eds.). *Science and drama: Contemporary and creative approaches to teaching and learning* (pp. 145–161). https://doi.org/10.1007/978-3-030-84401-1_9

Russon, J. (1994). Embodiment and responsibility: Merleau-Ponty and the ontology of nature. In Pyeonghwa K. (Ed.), *Man and world* (Vol. 27, pp. 291–308). Kluwer Academic.

Saricayir, H. (2010). Teaching electrolysis of water through drama. *Journal of Baltic Science Education, 9*(3), 179–168.

Swanson, C. J. (2021). Does being positioned in an expert scientist role enhance 11–13 year-old students' perceptions of themselves as scientists?. In P. J. White, J. Raphael, & K. van Cuylenburg (Eds.). *Science and drama: Contemporary and creative approaches to teaching and learning* (pp. 211–225). Springer. https://doi.org/10.1007/978-3-030-84401-1_13

Sumara, D. J., & Davis, B. (1997). Enlarging the space of the possible: Complexity, complicity and action-research practices. *Counterpoints, 67*, 299–312. http://www.jstor.org/stable/42975255

Tarlington, C., & Verriour, P. (1991). *Role drama.* Pembroke.

van Manen, M. (1990). *Researching lived experience: Human science for an action sensitive pedagogy.* Althouse Press.

Varela, F. (1987). Laying down a path in walking. In W. I. Thompson (Ed.), *GAIA, A way of knowing: Political implications of the new biology* (pp. 48–64). Lindisfarne Press.

Varela, F., Thompson, E., & Rosch, E. (1993). *The embodied mind: Cognitive science and human experience.* MIT Press.

Waldrop, M. M. (1992). *Complexity: The emerging science at the edge of order and chaos*. Simon & Schuster.

White, P. J., Raphael, J. & van Cuylenburg, K. (Eds.). (2021). *Science and drama: Contemporary and creative approaches to teaching and learning*. Springer.

Lynn Fels BA Honours (Queen's University, Theatre), BC Teaching Certificate (University of British Columbia), MA (Carleton University, Canadian Studies), PhD (University of British Columbia, Education) is a writer and professor of arts education in the Faculty of Education at Simon Fraser University. She was a researcher for Petro-Canada, humour columnist, and freelance arts educator. Returning to doctoral studies, she investigated storytelling and drama in science education with Karen Meyer. From 2002–2010, she was academic editor of *Educational Insights*, one of Canada's pioneering on-line educational research journals. Her research and teaching focuses on arts for social change, performative inquiry, mentorship *as performance*, teacher education, and performative writing. She delights in curricular interruptions and collaborative play with students and colleagues.

Karen Meyer BA (San Diego State University, Liberal Studies), MS (San Diego State University, Psychology and Natural Science Interdisciplinary Studies), PhD (University of British Columbia, Education) holds Emeritus standing after 26 years at the University of British Columbia in the Department of Curriculum and Pedagogy. Much of her career was spent in Teacher Education. She supervised 11 master's cohorts in Urban Education with practicing teachers (1998–2018). Another research and writing interest has been refugee education. As part of an international teaching and research project, she taught teachers in Dadaab Refugee Camp in northeastern Kenya, one of the largest camps in the world. Currently in retirement, she is a fiction writer and poet who volunteers in elementary classrooms as a resident writer.

Chapter 5
Teaching the Engineering Design Process: Preservice Teachers' Professional Development in a Community of Practice

Dawn L. Sutherland

5.1 The Engineering Design Process in Elementary Schools

In Canada, national curricular reform in science was initiated when the Council of Ministers of Education, Canada published the *Common Framework of Science Learning Outcomes, K to 12* (Council of Ministers of Education, 1997). The Framework emphasized the role of science, technology, society, and the environment (STSE) in science literacy; it was the foundation document used for science curriculum development in most Canadian provinces. Since 1998, when the first revised curriculum document in Canada was produced by the Ontario Ministry of Education, all subsequent provincial curriculum documents have included aspects of science and technology. The K–4 and Grades 5–8 science curriculum documents in Manitoba include both science and technology (Manitoba Education and Training, 1999, 2000). The term *technology* in these curricula typically refers to engineering and the design process.

The engineering design process consists of an open-ended, problem-based challenge that leaves ample opportunity for individual creativity (Hynes, 2010). The United States National Research Council (2011) characterized the engineering design process as the following: (a) a systematic process for solving engineering problems based on scientific knowledge and models of the material world; (b) proposed solutions result from a process of balancing competing criteria of desired functions, technological feasibility, cost, safety, esthetics, and compliance with legal requirements; and (c) there can be a range of solutions. Recent research suggests that engineering education can support children to develop problem-solving skills (English & Moore, 2018), develop the engineering and technological literacy skills needed for the

D. L. Sutherland (✉)
University of Manitoba, Winnipeg, MB R3T 2N2, Canada
e-mail: dawn.sutherland@umanitoba.ca

© The Author(s), under exclusive license to Springer Nature Switzerland AG 2023
C. D. Tippett and T. M. Milford (eds.), *Exploring Elementary Science Teaching and Learning in Canada*, Contemporary Trends and Issues in Science Education 53,
https://doi.org/10.1007/978-3-031-23936-6_5

twenty-first century (Cunningham et al., 2018), and improve their ability in science and mathematics (Cunningham et al., 2020).

Despite Canada's focus on STSE in curriculum documents, teaching the engineering design process—especially at the elementary level—is not well understood. Further, it can be challenging for teacher educators to provide sound, evidence-based pedagogical suggestions for engineering design to future teachers (Pendergast et al., 2011). These challenges suggest a need to research the teaching of engineering design and problem solving in the classroom. Research studies that provide suggestions for effective teaching of the design process at the elementary level are limited to date (Dubosarsky et al., 2018; English & Moore, 2018; Purzer & Douglas, 2018). One example of teaching engineering design to young children was an examination of Grades 1–4 students' use of scale model drawings to scaffold from design ideas to the product found that, for this age group, discussion and brainstorming were more useful in planning (Hill & Anning, 2001). Other studies have investigated young children's mechanistic reasoning in creating models to explain the motion of simple mechanical systems (Bolger et al., 2012; Hill, 1998). However, further research on effective ways to teach the engineering design process and engineering problem solving at the elementary level is sorely needed.

There have been several research studies on professional development in K–12 engineering education. This body of research tends to originate from the few jurisdictions such as Canada, New Zealand, Australia and the UK that have engineering education integrated into their curriculum (Mesutoglu & Baran, 2021). Most Canadian provinces have engineering education in their elementary curricula because the technological design process was included as a foundation process in the *Common Framework of Science Learning Outcomes for K–12* (Council of Ministers of Education, Canada, 1997) as an attempt to encourage a national curriculum. Professional development in engineering education benefits teachers in their ability to integrate engineering concepts and practices (Avery & Reeve, 2013; Baker et al., 2007; Hynes, 2012; Ross et al., 2016; Winn et al., 2009; Yoon et al., 2013) and to improve and instill more positive beliefs and attitudes towards engineering instruction (Autenrieth et al., 2017; Guzey et al., 2014; Hardré et al., 2017; Hynes, 2012; Ross et al., 2016; Utley et al., 2019; Yoon et al., 2013).

Collaboration with peers and experts as professional learning communities contributes to teacher professional growth (Baker et al., 2007; Duncan et al., 2011; Guzey et al., 2014; Hardré et al., 2010). Some ways in which professional development is situated is by creating collaborative teams that include university faculty, engineering students and working engineers.

This chapter describes the process a group of preservice teachers used to examine a variety of teaching strategies while working for an after-school and summer bridging program called *Design It*. The preservice teachers became a community of practice and collaborated on the design and implementation of a variety of design-based challenges and explored effective teaching practices using a collaborative inquiry process. This chapter explores the reflections of these preservice teachers, who are now teachers, on how their participation in *Design It* impacted their current classroom practices.

5.1.1 Design It: A Manitoba-based Engineering Design Process Program

Although engineering design and problem-solving have been in the Manitoba science curriculum documents (Manitoba Education, 1999, 2000) for Kindergarten through Grade 8 since 2000, engineering design education in K–12 has really only become of interest in the past 5 years. For example, of the 734 articles published in the *Canadian Journal for Science, Mathematics, and Technology Education* from 2001 to 2021, only 14% touch on engineering education in K–12 classrooms. Breaking this number down further, 12% of the 2001–2015 articles compared with 19% from 2016 to 2021 mention engineering education in K–12 classrooms. In response to this lack of attention, *Design It* was created in 2009, by the author, as an informal science education initiative to introduce inner-city youth to the engineering design process (Sutherland, 2019). The *Design It* program, led and supported by the author, worked closely with the local Boys and Girls Club of Winnipeg and three school divisions to offer engineering design in their summer bridging and after-school programs. University preservice teachers taught design-based problem solving in these programs. As a result, from 2009 to 2019 *Design It* created over 15 different design lessons and workshops as well as offered a 1-week intensive summer camp. Each year, the *Design It* team introduced over 1,500 students to engineering design problem-solving activities.

The preservice teachers hired to be the instructors for *Design It* attended an introductory workshop, taught by the author, focused on creating engineering lessons that emphasized initiating problem-based learning through literature and/or real-world problems and supporting planning skills in young children. Then the teaching team worked collaboratively to create a set of engineering design lessons that incorporated these principles. The *Design It* lessons always included a literacy component to create the context, instruction on the science concepts involved in the problem, time for planning and some instruction on technical drawing, an opportunity to build and test designs, and a discussion that included a reflection on the process and on the way design thinking works.

5.1.2 Professional Development of Preservice Teachers Through Collaborative Inquiry

Design It, however, was not only a way to introduce Manitoba inner-city youth to the engineering design process; it was also a vehicle for the professional development of future elementary teachers who may not have the self-confidence to teach science or engineering design (Bencze, 2010; Park et al., 2017). Student teachers are often not exposed to teaching the engineering design process in practicum placements because it is not a priority for in-service teachers (Bencze, 2010). Being a *Design It* instructor was a way for preservice teachers to develop self-efficacy in teaching engineering design and open-ended problem solving in the classroom.

When preservice teachers participated as instructors in *Design It*, they joined a community of practice that supported and strengthened their teaching skills in the design process. The professional development opportunities provided by *Design It* included the creation of a community of practice to brainstorm possible design challenges for the workshops, create lessons in small groups, teach the lessons to the rest of the team, and critique and refine lessons before offering them to students. Through this collaborative process, the preservice teachers developed a collective responsibility (Whalan, 2012) toward the teaching of engineering design and problem-solving skills to inner-city youth.

5.1.3   Collaborative Inquiry: Furthering the Professional Development of Preservice Teachers in Engineering Design

Over the past 15 years, collaborative inquiry (CI) has emerged as a dominant framework for educator classroom-based research (Butler & Schnellert, 2012; DeLuca et al., 2015; Lehman et al., 2014; McGarr et al., 2019; Nelson et al., 2010). Buschor and Kamm (2015) stated that "the goal of collaborative inquiry is to move towards a system of problem-solving in communities of practice that promote[s] the development of knowledge in the practitioners' context" (p. 234). The process of collaboration in CI is a particular type of social interaction where educators investigate focused aspects of their professional practice by exploring and documenting student response to instruction, leading to new understandings and responsive actions (Lee, 2009). Several research programs have explored the structure of the collaborative process (Butler & Schnellert, 2012; Nelson & Slavit, 2008; Nelson et al., 2010). Regardless of the particular CI process, a common theme is dialogical sharing. Through dialogical sharing, teachers use their individual knowledge as the basis for co-constructing deeper, shared knowledge (Kennedy et al., 2011; Nelson et al., 2010). Some studies have examined different methods of dialogical sharing, such as shared participation or the construction of a shared vision (Nelson & Slavit, 2008), as well as the use of documentation as a tool to facilitate dialogue (Given et al., 2009). Many CI initiatives have been in school settings with practicing teachers; however, there is growing evidence that the CI process is effective in preservice teacher education (McGarr et al., 2019; Willegems et al., 2017).

At the direction of the author, for each lesson and workshop the *Design It* team created, the preservice teachers collaboratively created an inquiry question to explore different aspects of teaching the design process. Although there were many versions of the engineering design process in curriculum and engineering education resources, at the time there was not a great deal of research on what comprised a good design lesson. More recently, researchers have begun to explore the parameters in engineering activity and design that are foundational engineering concepts and practices (Cunningham et al., 2018). The *Design It* team conducted inquiries that explored effective engineering pedagogy. For example, prior to teaching a lesson on porosity and fishing net design to Grade 2 students, the instructors decided to explore the question, "If we provide some direct instruction on basic ideas in technical drawing, does this impact the way in which the students engage with the design itself?" The instructors then co-created a component on technical drawing to be included in the lesson, taught it, observed and journaled about the experience, then discussed their findings with the group. A second example is a lesson on designing a water filtration system; after teaching this lesson a few times, the instructors thought that the students did not have a clear grasp of how different materials absorb water. They then developed an inquiry to see if the inclusion of a short science inquiry on how different materials absorb water affected the design of the filtration device and met to discuss their findings. A similar process was used for each lesson and workshop in *Design It*. The following study is an exploration of the impact of the professional development opportunities inherent in *Design It*, as perceived by preservice teachers who are currently classroom teachers.

5.2   Method

This study took place 2 years after the last *Design It* workshop was offered to school divisions. The study participants were six *Design It* instructors who went on to become teachers in the public school system. Table 5.1 contains information about the participants' characteristics, including number of years in *Design It*, graduation year, stream, major subject focus, and current teaching position.

The *Design It* program required the instructional team, consisting of 6–14 preservice teachers depending on the year, to work collaboratively in weekly meetings to create, plan, and implement engineering design lessons for inner-city youth in Winnipeg, Manitoba. The collaborative process began with the instructional team

Table 5.1 Characteristics of former *Design It* instructors

Participant (pseudonym)	Years in *Design It*	Year graduated	Stream	Major	Years teaching	Grade level
Quin	2	2016	Middle	Mathematics	4	K–6
Lynn	3	2018	Elementary	English	2	1–2
Dale	3	2018	Elementary	English	2	4
Jill	4	2016	Senior	Biology	4	7–8
Lindsay	4	2011	Middle	English	7	K–6
James	6	2016	Senior	Biology, Chemistry	4	9–12

conducting a brainstorming activity to identify possible engineering design challenges. Through discussion, debate, and a vote, the brainstormed list was narrowed to approximately eight challenges that could be offered throughout the year. The preservice teachers worked in pairs to design lessons for two of the challenges. Each week, one pair would present their lesson plan to the community of practice and subsequently lead them through the lesson. The group would offer feedback, and the community of practice would collaborate to develop a final lesson plan for that particular engineering challenge. Once the lesson was finalized, the group would create an inquiry-based pedagogical question that would be explored during the instructional time. All members of the instructional team would then go and teach the lesson to various classroom and after-school sites for a 2-week period. After each lesson implementation, the team would meet to report, reflect, and further refine the plan for future iterations. All lessons and workshops in *Design It* were posted to the website (https://dsutherlan4.wixsite.com/designit) for classroom teachers and informal science educators.

This study was approved by the Psychology-Sociology Research Ethics Board at the University of Manitoba and incorporated a qualitative descriptive methodology to examine the issue following Sandelowski (2000) and Braun and Clarke (2006) in employing thematic analysis across the data rather than simply within individual items. As part of the study, the six teachers participated in a 60–90-min semi-structured interview in which they were asked to reflect on their experiences in *Design It* and asked how those experiences impacted their transition from preservice to in-service teacher.

A semi-structured interview guide was used to ensure consistency in data collection. Questions were developed that addressed planning (e.g., How do you plan your lessons and develop instructional strategies in your past and current schools? What might be your preferred way to plan-in an ideal setting?), critical incidents (e.g., Describe some incidents from teaching with *Design It* that you remember that you feel may have impacted your current practices as a teacher. Why do you think these incidences stand out?), and skills (e.g., What teaching skills do you think you developed while participating in *Design It*?). With the participants' permission, all interviews were recorded and then transcribed verbatim to ensure accuracy and reliability. Each participant was sent the interview transcript for information and approval. The study incorporated several analytical methods including thematic analysis employing deductive coding (Braun & Clarke, 2006; Evans & Lewis, 2018) and thick description (Creswell & Miller, 2000).

5.3   Results

The interview transcripts captured the participants' rich reflections on their *Design It* experiences and the impact the program had on their professional development. Three primary themes emerged in the analysis of the transcripts: planning, confidence, and problem solving as an instructional strategy. Note that pseudonyms are used for the participants and that their responses have been edited for clarity.

5.3.1 Planning

All participants spoke about the value of the collaborative planning process that was at the heart of the *Design It* program. Participants highlighted three key aspects: planning as a team, gathering multiple points of view about a lesson, and planning strategies.

The collaborative process helped participants anticipate possible challenges and identify areas for possible pedagogical inquiry. For example, Jill described the process as

> really interesting because we all came from different backgrounds and I think that was part of the point of it; there was math people, science people, English people and we would just come with all our ideas and we would tap into certain strengths, like finding a literacy person for providing a book, or who's really great at hands-on science and can take the lead on this one. So it was just very collaborative, we would finalize an idea, whoever's idea it was they would,— I think they were in charge of creating a lesson plan for it, or a design brief is what we called it, and we would try them out in the office, we would have … our own building day where we tried out the prototypes and then from there we would kind of go, okay, so what problems can we foresee happening with the kids?

One notable impact associated with the community of practice was team planning. All participants viewed the team planning aspect as beneficial, especially from the point of view of their position as new teachers who were experiencing how isolated the planning process can be. For example, Dale stated,

> It was nice to just be able to bounce ideas off of each other to create those new lesson plans … yeah, just working together to see what that would look like and, yeah, mostly I like the idea of just being able to share ideas with one another 'cause now teaching by myself, sometimes it's [*chuckle*] hard to think of all the different things that you can do, right?

Lindsay reflected,

> People plan on their own … and go through units on their own and it can be kind of isolating.… In retrospect, I can see how that group planning and sharing of expertise opened me up to planning more now with my fellow colleagues, and I really value that type of collaboration. And … it was different than group work we did in university, like it was somehow more … meaningful and purposeful, … probably because we were actually working with students.

The participants identified the value in discussing ideas with a diverse group of teachers who had multiple points of view. This value was reflected in comments such as James's response:

> I got to work with different teachers that have different backgrounds, so they always add input on what are ways to present a scenario, … so going in to a professional teaching job … with the collaborative experience … really helped me … communicate my ideas as well as being able to be open to other ideas that are not necessarily scientific per se.

Some participants identified the value of collaborating with individuals who had a science background. This shared expertise allowed less experienced participants to learn to examine the design problems from a more scientific point of view. For example, Lindsay remarked,

> that was interesting too because they both had Science backgrounds … so I remember some of … their ideas being a bit more complex … since I don't really have a science background…. It was a good mix 'cause I was able to see it from the students' perspective a bit more and … it was relatively new for me.

The teachers identified how the collaborative environment helped them learn and develop planning strategies as they co-constructed lessons. Lynn identified the planning process as beginning with

> a big idea and then … scaffolding it down into … what the end goal was, what the end goal we wanted was…. I still plan like I'm in *Design It* I think, … I still always start with a big idea and or start with a main idea and then after looking at the outcomes I need to teach, … how can I get kids to really … think about this and really … have it be meaningful to them, and so I think that when I'm planning now I kind of think about all those things, so … I'm not so narrow minded.

5.3.2 Confidence

The second theme that emerged from the thematic analysis was ways in which teaching for *Design It* impacted the participants' confidence. Participants highlighted three key aspects: teaching engineering design, teaching in general, and teaching science.

Participants described how they had become more confident in teaching the engineering design process because they had observed the multiple ways in which it could play out in the classroom. For example, Jill said, "well, I'm definitely more confident … doing *Design It* … design challenges in my classroom because I've had firsthand experience of how to plan and how to do them and how to deliver those lessons."

Participants reflected on teaching in general and how the *Design It* experience had helped them to develop more confidence in the classroom. James reflected on how they found their teacher voice while working with *Design It*:

> I had trouble having my own teacher voice…. I remember 'P' even though she was not— she wasn't loud, she knew how to command a classroom by being gentle and so that in itself was a skill that I needed … to really learn how to do … 'cause it's … not natural.

Lynn identified how teaching with *Design It* helped her become less reliant on the lesson plan:

> I'll never forget I was … studying this lesson plan … and I was so nervous to teach it 'cause I was like, well, what if I don't ask the right question, or what if I don't have the right answer or, and I'll never forget feeling that … nervousness and then now, if I was to teach a *Design It* lesson,… I could do the whole lesson without reading this piece of paper and it isn't because you memorized it but it's because … you just know what you're doing.

Participants focused on how their confidence in their ability to teach science had increased. Three of the six participants did not have science backgrounds when they were hired to teach for the *Design It* program. This choice was intentional as most teachers at the elementary level do not have strong science backgrounds and so teaching the engineering design process was something that needed to be learned. Dale commented,

> I remember ... I never used to consider myself a science-y person and then ... getting into DI and just doing lessons that were specific to science in such a fun way that was so engaging for the kids and had such a positive lessons for the kids, it just changed my whole perspective ... on science in particular, of course, but also just ... in teaching in general.... like asking open questions and just encouraging them that way.... I know ... I've brought it into many more subjects than just science.

Lindsay, who did not have a science background, found themself mentoring teachers at their school early in their career.

> I found other teachers, especially older teachers, would really shy away from the design outcomes and I felt so familiar with that ... that part of the curriculum really energized me ... it was so hands on and we're ... combining Science and Social Studies outcomes, ... as a teacher you really had to think hard about ... what the essential, essential understandings are and ... the bigger outcomes you want, so you can combine those things so your students have ... a very rich integrative learning experience. So I found my background and familiarity with the design process for Science easily transferred to teaching grade eight curriculum and that made me ... really excited about teaching Science.

5.3.3 Problem Solving as an Instructional Strategy

The third theme that emerged from the thematic analysis was problem solving as an instructional strategy. Participants highlighted two key aspects: making real-world connections and providing flexible support.

Participants identified how the *Design It* team sought engineering problems that contained a real-world connection. For example, Lynn emphasized possibilities for developing empathy through real-world examples, reflecting,

> I remember when the Shoal Lake issue was ... a really big concern ... and a lot of the schools wanted to implement that into their curriculum ... for their students to be aware and so we kind of took that and said, ... how can we educate kids on it and actually make them empathetic about the situation.

Participants described how they learned to flexibly support students during open-ended problem solving while working with *Design It*. Lindsay described supporting children when they experienced frustration:

> one of the things ... would happen every time, every single time we did a challenge and students want to give up because they didn't get the outcome they expected or ... maybe it's a little different than what they're used to or there's ... obviously a process you need to follow and maybe they're not comfortable with that, but it certainly helped me to think creatively on how to encourage students to not give up and to persevere and to keep going or

try something new. And even now when I'm in the classroom, I use those creative thinking skills and try to think outside of the box for ways to engage them and encourage them to think differently about a problem or try again in a different way or, yeah, think more … critically about what they're working on.

It was clear from the interviews that participating in *Design It* was a positive experience for each instructor. Even two years after the last workshop, they were able to clearly identify areas in which participating had impacted their pedagogy and confidence in teaching the design process and problem solving. Planning collaboratively, anticipating challenges, being flexible, creatively introducing an engineering problem attached to real-world issues, and feeling confident in facilitating an open-ended problem-solving lesson to a multitude of students are all skills that *Design It* fostered in these future teachers of young children.

5.4 Discussion

Although science and technology curricula in Canada contain engineering design as a consideration in science education, how to teach engineering design is not well understood. Only recently has research in engineering teaching at the K–12 level focused on aspects such as beliefs and readiness (Park et al., 2017), trajectories in engineering education (Cunningham et al., 2018, 2020; Lachapelle et al., 2019), and challenges associated with science, technology, engineering, and mathematics (STEM) reform (Bernstein-Sierra & Kezar, 2017; Murray, 2019).

Teacher beliefs about and readiness to teach the engineering design process are factors that contribute to the implementation of the design process in the classroom, particularly in early childhood (Park et al., 2017). In the early years, teacher readiness is related not only to an increase in experience but also an awareness of the challenges that may be encountered when teaching the design process (Park et al., 2017). Research has shown that creating a community of practice, focused on engineering design, among university faculty and elementary teachers can positively impact teacher beliefs and confidence (Lehman et al., 2014). It was clear from the reflections of these *Design It* instructors that their experience increased their confidence in teaching engineering and science because they could respond to the different ways design can "play out" in the classroom. The research described here provides insights into how a community of practice composed primarily of preservice teachers teaching the design process to inner-city youth can have long lasting impacts. The teacher–instructors in *Design It* emphasized all the elements that created a strong community of practice such as mutual respect, opportunities to share ideas, and an environment to apply knowledge (Li et al., 2009).

The collaborative planning and inquiry experience of participating in *Design It* helped the instructors understand the possible challenges younger students may encounter while working through an engineering problem. Teaching the same design lesson several times to a variety of students—something that rarely happens in elementary classrooms—provided an opportunity for the instructors to conduct some deeper pedagogical inquiries. They began to ask questions about how to teach engineering problems effectively through a CI process that contributed to their individual growth in planning and confidence in teaching the design process. This CI process was a way to explore foundational engineering concepts and practices and a developmental trajectory for engineering skills. Recent research supports a developmental perspective on teaching engineering skills. For example, in the earlier grades the properties of materials, how they interact, and the design process have been suggested as a set of design parameters most suitable for primary students (Cunningham et al., 2018, 2020).

Student teachers are often not exposed to teaching the engineering design process in practicum placements because it is not a priority for in-service teachers (Bencze, 2010). This omission creates a challenge in reforming and changing the way the engineering design process is taught to young children. In their analysis of best change strategies used to promote change in STEM instructional practices, Henderson et al. (2011) identified four common strategies: dissemination of curriculum and pedagogy, developing reflective teachers, enacting policy, and developing shared vision. Only long-term strategies that were aligned with or seek to change the beliefs of teachers, such as reflection and creating shared visions, were found to be effective strategies. This point is important when considering any reform to preservice teacher education and teacher professional development in engineering education. The reflections of the teachers in *Design It* two years later highlighted how the community of practice was important in the creation of a shared vision on what constitutes a design lesson and reflection on how to refine and enhance engineering instruction. These findings support the suggestion of Henderson et al. (2011) about how to move toward change in engineering education in the elementary years.

Although the CI in *Design It* emerged at the time as an important component to planning and design, it does not seem to have become an aspect of these teachers' current reflective practice; at least it was not mentioned in these interviews. This finding seems to be in congruence with other research around the effectiveness of peer-discussions in CI. Bushchor and Kamm (2015) stated that "The goal of collaborative inquiry is to move towards a system of problem-solving in communities of practice that promote the development of knowledge in the practitioners' context" (p. 234). Inclusion of CI within the peer community of practice supports some recent findings on improving the reflective practice of preservice teachers in STEM classrooms (McGarr et al., 2019). The reflections of these *Design It* instructors certainly highlighted the importance of the collaborative process but do not elaborate on the pedagogical inquiries that were conducted within the community of practice, nor do they identify inquiry as a practice in their current planning or instruction. This finding is congruent with those of McGarr et al. (2019) where peer-discussions on a classroom critical incident supported identifying and questioning some preconceived assumptions about teaching but did not support greater depth in reflective thinking. They suggested that for critical reflection preservice teachers need to engage with experienced teachers.

Canada is not the only country that includes the engineering design process in its curriculum; for example, Australia and United Kingdom also focus on design.

Therefore, these findings will be relevant for an international audience of science and technology teachers, school administrators, and informal educators. The study results contribute to an improved understanding of how a community of practice can support teachers in learning how to teach engineering design.

References

Autenrieth, R. L., Lewis, C. W., & Butler-Purry, K. L. (2017). Long-term impact of the enrichment experiences in engineering (E^3) summer teacher program. *Journal of STEM Education, 18*(1), 25–31.
Avery, Z. K., & Reeve, E. M. (2013). Developing effective STEM professional development programs. *Journal of Technology Education, 25*(1), 55–69.
Baker, D., Yasur-Purzer, S., Kurpius, S. R., Krause, S., & Roberts, C. (2007). Infusing design, engineering and technology into K-12 teachers' practice. *International Journal of Engineering Education, 23*(5), 884–893.
Bencze, J. L. (2010). Promoting student-led science and technology projects in elementary teacher education: Entry into core pedagogical practices through technological design. *International Journal of Technology and Design Education, 20*(1), 43–62. https://doi.org/10.1007/s10798-008-9063-7
Bernstein-Sierra, S., & Kezar, A. (2017). Identifying and overcoming challenges in STEM reform: A study of four national STEM reform communities of practice. *Innovative Higher Education, 42*(5), 407–420. https://doi.org/10.1007/s10755-017-9395-x
Bolger, M. S., Kobiela, M., Weinberg, P. J., & Lehrer, R. (2012). Children's mechanistic reasoning. *Cognition and Instruction, 30*(2), 170–206. https://doi.org/10.1080/07370008.2012.661815
Braun, V., & Clarke, V. (2006). Using thematic analysis in psychology. *Qualitative Research in Psychology, 3*(2), 77–101. https://doi.org/10.1191/1478088706qp063oa
Buschor, C. B., & Kamm, E. (2015). Supporting student teachers' reflective attitude and research-oriented stance. *Educational Research for Policy and Practice, 14*(3), 231–245. https://doi.org/10.1007/s10671-015-9186-z
Butler, D. L., & Schnellert, L. (2012). Collaborative inquiry in teacher professional development. *Teaching and Teacher Education, 28*(8), 1206–1220. https://doi.org/10.1016/j.tate.2012.07.009
Council of Ministers of Education, Canada. (1997). *Common framework of science learning outcomes, K to 12: Pan-Canadian protocol for collaboration on school curriculum for use by curriculum developers*. Toronto, ON: Author. https://archive.org/details/commonframework00coun
Creswell, J. W., & Miller, D. L. (2000). Determining validity in qualitative inquiry. *Theory into Practice, 39*(3), 124–130. https://doi.org/10.1207/s15430421tip3903_2
Cunningham, C. M., Lachapelle, C. P., & Davis, M. E. (2018). Engineering concepts, practices, and trajectories for early childhood education. In L. English & T. Moore (Eds.), *Early engineering learning* (pp. 135–174). Springer. https://doi.org/10.1007/978-981-10-8621-2_8
Cunningham, C. M., Lachapelle, C. P., Brennan, R. T., Kelly, G. J., Tunis, C. S. A., & Gentry, C. A. (2020). The impact of engineering curriculum design principles on elementary students' engineering and science learning. *Journal of Research in Science Teaching, 57*(3), 423–453. https://doi.org/10.1002/tea.21601
DeLuca, C., Shulha, J., Luhanga, U., Shulha, L. M., Christou, T. M., & Klinger, D. A. (2015). Collaborative inquiry as a professional learning structure for educators: A scoping review. *Professional Development in Education, 41*(4), 640–670. https://doi.org/10.1080/19415257.2014.933120
Dubosarsky, M., John, M. S., Anggoro, F., Wunnava, S., & Celik, U. (2018). Seeds of STEM: The development of a problem-based STEM curriculum for early childhood classrooms. In L.

English & T. Moore (Eds.), *Early engineering learning* (pp. 249–269). Springer. https://doi.org/10.1007/978-981-10-8621-2_12

Duncan, D., Diefes-dux, H., & Gentry, M. (2011). Professional development through engineering academies: An examination of elementary teachers' recognition and understanding of engineering. *Journal of Engineering Education, 100*(3), 520–539. https://doi.org/10.1002/j.2168-9830.2011.tb00025.x

English, L., & Moore, T. (2018). Early engineering learning. *Springer*. https://doi.org/10.1007/978-981-10-8621-2

Evans, C., & Lewis, J. (2018). *Analysing semi-structured interviews using thematic analysis: Exploring voluntary civic participation among adults.* SAGE Publications Ltd.

Given, H., Kuh, L., LeeKeenan, D., Mardell, B., Redditt, S., & Twombly, S. (2009). Changing school culture: Using documentation to support collaborative inquiry. *Theory into Practice, 49*(1), 36–46. https://doi.org/10.1080/00405840903435733

Guzey, S. S., Tank, K., Wang, H.-H., Roehrig, G., & Moore, T. (2014). A high-quality professional development for teachers of Grades 3–6 for implementing engineering into classrooms: Engineering integration. *School Science and Mathematics, 114*(3), 139–149. https://doi.org/10.1111/ssm.12061

Hardré, P. L., Ling, C., Shehab, R. L., Nanny, M. A., Nollert, M. U., Refai, H., Ramseyer, C., Herron, J., Wollega, E. D., & Huang, S.-M. (2017). Situating teachers' developmental engineering experiences in an inquiry-based, laboratory learning environment. *Teacher Development, 21*(2), 243–268. https://doi.org/10.1080/13664530.2016.1224776

Hardré, P. L., Nanny, M., Refai, H., Ling, C., & Slater, J. (2010). Engineering a dynamic science learning environment for K-12 teachers. *Teacher Education Quarterly, 37*(2), 157–178.

Henderson, C., Beach, A., & Finkelstein, N. (2011). Facilitating change in undergraduate STEM instructional practices: An analytic review of the literature. *Journal of Research in Science Teaching, 48*(8), 952–984. https://doi.org/10.1002/tea.20439

Hill, A. M. (1998). Problem solving in real-life contexts: An alternative for design in technology education. *International Journal of Technology and Design Education, 8*(3), 203–220. https://doi.org/10.1023/A:1008854926028

Hill, A. M., & Anning, A. (2001). Primary teachers' and students' understanding of school situated design in Canada and England. *Research in Science Education, 31*(1), 117–135. https://doi.org/10.1023/A:1012662329259

Hynes, M. M. (2010). Middle-school teachers' understanding and teaching of the engineering design process: A look at subject matter and pedagogical content knowledge. *International Journal of Technology and Design Education, 22*(3), 345–360. https://doi.org/10.1007/s10798-010-9142-4

Hynes, M. M. (2012). Middle-school teachers' understanding and teaching of the engineering design process: A look at subject matter and pedagogical content knowledge. *International Journal of Technology and Design Education, 22*(3), 345–360. https://doi.org/10.1007/s10798-010-9142-4

Kennedy, A., Deuel, A., Nelson, T. H., & Slavit, D. (2011). Requiring collaboration or distributing leadership? *Phi Delta Kappan, 92*(8), 20–24. https://doi.org/10.1177/003172171109200805

Lachapelle, C. P., Cunningham, C. M., & Oh, Y. (2019). What is technology? Development and evaluation of a simple instrument for measuring children's conceptions of technology. *International Journal of Science Education, 41*(2), 188–209. https://doi.org/10.1080/09500693.2018.1545101

Lee, T. (2009, May). *Educational leadership for the 21st century: Leading school improvement through collaborative inquiry* [Paper presentation]. Canadian Association for Studies in Educational Administration conference, Ottawa, ON, Canada.

Lehman, J. D., Kim, WooRi, & Harris, C. (2014). Collaborations in a community of practice working to integrate engineering design in elementary science education. *Journal of STEM Education: Innovations & Research, 15*(3), 21–28.

Li, L. C., Grimshaw, J. M., Nielsen, C., Judd, M., Coyte, P. C., & Graham, I. D. (2009). Evolutions of Wenger's concept of community of practice. *Implementation Science, 4*(1), Article 11. https://doi.org/10.1186/1748-5908-4-11

Manitoba Education and Training. (1999). *Kindergarten to Grade 4 science: Manitoba curriculum framework.*

Manitoba Education and Training. (2000). *Grade 5–8 science: Manitoba curriculum framework.*

McGarr, O., McCormack, O., & Comerford, J. (2019). Peer-supported collaborative inquiry in teacher education: Exploring the influence of peer discussions on pre-service teachers' levels of critical reflection. *Irish Educational Studies, 38*(2), 245–261. https://doi.org/10.1080/03323315.2019.1576536

Mesutoglu, C., & Baran, E. (2021). Integration of engineering into K-12 education: A systematic review of teacher professional development programs. *Research in Science & Technological Education, 39*(3), 328–346. https://doi.org/10.1080/02635143.2020.1740669

Murray, J. (2019). Routes to STEM: Nurturing science, technology, engineering and mathematics in early years education. *International Journal of Early Years Education, 27*(3), 219–221. https://doi.org/10.1080/09669760.2019.1653508

Nelson, T. H., Deuel, A., Slavit, D., & Kennedy, A. (2010). Leading deep conversations in collaborative inquiry groups. *The Clearing House: A Journal of Educational Strategies, Issues and Ideas, 83*(5), 175–179. https://doi.org/10.1080/00098650903505498

Nelson, T., & Slavit, D. (2008). Supported teacher collaborative inquiry. *Teacher Education Quarterly, 35*(1), 99–116. http://www.jstor.org/stable/23479033

Park, M.-H., Dimitrov, D. M., Patterson, L. G., & Park, D.-Y. (2017). Early childhood teachers' beliefs about readiness for teaching science, technology, engineering, and mathematics. *Journal of Early Childhood Research, 15*(3), 275–291. https://doi.org/10.1177/1476718X15614040

Pendergast, D., Garvis, S., & Keogh, J. (2011). Pre-service student-teacher self-efficacy beliefs: An insight into the making of teachers. *Australian Journal of Teacher Education, 36*(12), Article 4. https://doi.org/10.14221/ajte.2011v36n12.6

Purzer, Ş., & Douglas, K. A. (2018). Assessing early engineering thinking and design competencies in the classroom. In L. English & T. Moore (Eds.), *Early engineering learning* (pp. 113–132). Springer. https://doi.org/10.1007/978-981-10-8621-2_7

Ross, J. M., Jackson-Lee, Y., & Singer, J. E. (2016). Professional development for the integration of engineering in high school STEM classrooms. *Journal of Pre-College Engineering Education Research (J-PEER), 6*(1), 1–16. https://doi.org/10.7771/2157-9288.1130

Sandelowski, M. (2000). Whatever happened to qualitative description? *Research in Nursing & Health, 23*(4), 334–340. https://doi.org/10.1002/1098-240X(200008)23:4%3c334::AID-NUR9%3e3.0.CO;2-G

Sutherland, D. (2019). Science education in Manitoba: Collaborative professional communities. In C. D. Tippett & T. M. Milford (Eds.), *Science education in Canada* (pp. 85–102). Springer. https://doi-org.uml.idm.oclc.org/10.1007/978-3-030-06191-3_5

United States National Research Council. (2011). *Successful K-12 STEM education: Identifying effective approaches in science, technology, engineering, and mathematics.* https://doi.org/10.17226/13158

Utley, J., Ivey, T., Hammack, R., & High, K. (2019). Enhancing engineering education in the elementary school. *School Science and Mathematics, 119*(4), 203–212. https://doi.org/10.1111/ssm.12332

Whalan, F. (2012). *Collective responsibility: A redefining what falls between the cracks for school reform* (1st ed.), Brill, Sense. https://doi.org/10.1007/978-94-6091-882-7

Willegems, V., Consuegra, E., Struyven, K., & Engels, N. (2017). Teachers and pre-service teachers as partners in collaborative teacher research: A systematic literature review. *Teaching and Teacher Education, 64*, 230–245. https://doi.org/10.1016/j.tate.2017.02.014

Winn, G. L., Lewis, D., & Curtis, R. (2009). Bridging engineering education to high school science teachers using time kits. *The International Journal of Engineering Education, 25*(3), 493–498.

Yoon, S. Y., Diefes-Dux, H., & Strobel, J. (2013). First-year effects of an engineering professional development program on elementary teachers. *American Journal of Engineering Education, 4*(1), 67–84. https://doi.org/10.19030/ajee.v4i1.7859

Dawn L. Sutherland BSc Hons (Queen's University), MSc (University of Manitoba), PhD (University of Nottingham) is a professor, graduate Chair and department head in the Faculty of Education at the University of Manitoba. She was a Canada Research Chair in Indigenous Science Education in 2006 (renewed in 2011 as Canada Research Chair in Science Education in Cultural Contexts) and explores the relationship between culture and science education in Indigenous and inner-city communities. She is particularly interested in engineering design process and problem solving in cultural contexts. As well, Dr. Sutherland has worked with Dr. Kathyrn Levine for the past 20 years on resilience and career development in inner-city environments and currently is involved in research related to inter-professional collaboration and educating children in care.

Chapter 6
Is This a Course About Science? Tensions and Challenges with Engaging Preservice Elementary and Middle Years Teachers in Science Learning

Tim Molnar

6.1 Introduction

The intent of this work is to share my experience during the establishment, development, and implementation of a new course—*Is this a course about science?*—with the hope that it will aid other science educators. Although I had taught the course three times as of the writing of this chapter, by virtue of the concerns and challenges that have emerged for me, I consider the course to be in transition and development. The course is intended to support elementary and middle years preservice teachers in science during their first or second year of a 4-year direct entry program. The course title emerged during initial discussions with faculty colleagues about the apparent overlapping and intersecting motivations and purposes of the course.

Despite over 15 years' experience instructing university science methods courses for elementary, middle years, and secondary preservice teachers, and over 18 years' experience teaching science at the secondary level, the development of this new course posed challenges for me, both intellectually and practically. After teaching the course through three iterations, I continue to have a sense of unease and to question what is necessary, what is desirable, and how productive and meaningful engagement with preservice teachers of science might be realized.

In this chapter, I share some of my concerns about how to aid preservice science teachers in building their conceptual and procedural knowledge, developing their understanding of the nature of science, and strengthening their confidence and belief in their capacity as science teachers. Discussion focuses on the context, need, and motivation for a new course; recommendations for science teacher preparation; establishing learning outcomes and activities, and meeting those outcomes. My

T. Molnar (✉)
College of Education, University of Saskatchewan, Saskatoon, SK, Canada
e-mail: tim.molnar@usask.ca

© The Author(s), under exclusive license to Springer Nature Switzerland AG 2023
C. D. Tippett and T. M. Milford (eds.), *Exploring Elementary Science Teaching and Learning in Canada*, Contemporary Trends and Issues in Science Education 53,
https://doi.org/10.1007/978-3-031-23936-6_6

thinking is informed by my experience as the course instructor, discussions with colleagues concerning the course, and my opinions of student artefacts. I also draw on student feedback from course evaluations, discussions with my teaching assistants, and results of surveys undertaken by the University of Saskatchewan Centre for Teaching and Learning concerning the course research experience. I end the chapter with conclusions concerning my attempt to create a meaningful science teaching and learning experience for early-in-program preservice teachers of science.

6.2 What Was the Need for the New Course?

The need for a new course emerged because of changes in the University of Saskatchewan College of Education preservice teacher program that were enacted in response to various stakeholder concerns about science learning. These changes prompted the course developers, predominantly me, to consider what motivations would underlie and propel the course design, outcomes, and activities.

6.2.1 Motivations

The College's recent transition to a direct entry 4-year teacher education program and the need to respond to various stakeholder concerns about science learning prompted the need for a new course focused on science. The Saskatchewan Ministry of Education (SME) had concerns regarding the 2015 provincial Programme for International Student Assessment results in science and the perceived need for educators to involve learners in science inquiry. Both the SME and the University expressed concerns that the number of science credits taken by preservice early and middle years teachers at the College was often minimal. College of Education program advisors noted that preservice teachers, especially in the early years, tended to avoid mathematics and sciences courses because they were fearful of these subjects; this fear was seen as something to address. The College of Arts and Science had an overall ambition to support sciences in the K–12 system, because it wanted to reduce the number of postsecondary students who avoided the sciences. As well, there was a perception among some faculty members that students did not do well in mandatory first year science courses. Thus, there was a strong desire to support preservice teachers in their science learning. The College of Arts and Science agreed to relinquish 3 credit units to be utilized by the College of Education for a newly developed course focused on science (D. Wallin, 2019, "personal communication").

As the primary science educator in the College of Education in 2017, I was invited by the Associate Dean of Undergraduate Programming to discuss the nature and structure for a new course focused on science. Subsequently, a team was formed, consisting of the Associate Dean, an instructor in the College who was also the outreach education coordinator from the Canadian Light Source (Canada's national synchrotron light

source facility located on the grounds of the University of Saskatchewan, https://lights
ources.org/), and me. We met several times to discuss our motivations for designing
the course as well as its nature and possible outcomes. There was confusion regarding
whether the course was to be focused on science or science methods. For my part,
this confusion constituted a central tension as we considered the needs of various
stakeholders. If the course were to be a science course, then with what science
knowledge and practices should preservice teachers be engaged? If the course were
to be a science methods course, how would it integrate with the subsequent mandatory science methods course? Uncertainty remained, leaving me to ask our small
working group: *Is this a course about science?* Ultimately, that question became the
course title. Working through this uncertainty and attempting to address the various
stakeholders' interests, the following motivations for the course became evident:

- Moving beyond science content knowledge development
- Modeling science inquiry teaching approaches
- Addressing and reducing anxiety regarding science
- Addressing a lack of preservice teachers' data literacy
- Developing a sense of belonging and identity among preservice teachers as
  capable of engaging with science and as confident science educators
- Developing an informed understanding and appreciation concerning the nature of
  science knowledge and process
- Providing opportunities for authentic inquiry-based research.

While not an exhaustive list, these motivations provide insight into our thinking
concerning the general direction of the course. Relying on the observations and
experience of colleagues in the Colleges of Education and Arts and Science, I sought
to clarify what these motivations meant for planning the course.

6.3 Recommendations for Preservice Science Teacher Preparation

Recommendations expressed in the literature also influenced my thinking about
the learning outcomes and activities for this new course. These recommendations
involved attending to preservice science teachers' self-efficacy, supporting teacher
engagement in science inquiry, and investigations into the nature of science. There
was also a need to relate our course experience to the provincial science curricula,
given that many of the preservice teachers would be teaching within the province. In
what follows, I offer a commentary on each of these influences.

6.3.1   Science Teacher Self-efficacy

Bergman and Morphew (2015) and Valls-Bautista et al. (2021) have suggested that
the preparation of elementary teachers to successfully teach science remains a central
issue in science education. They list a variety of challenges and obstacles experienced
by preservice teachers that are interwoven with how they understand and view science
and science education; for example, feelings of unpreparedness and lack of competence with science are common. Such challenges are encountered with preservice
teachers in Canadian universities destined for early years and middle years teaching
where these students typically take one or two science method courses (Tippett &
Milford, 2019). While such courses may be sufficient in initially developing instructional strategies, the science knowledge (i.e., content and process) that preservice
elementary and middle years teachers possess typically relies on their K–12 student
experience and their few university science courses.

Pajares (1992, 1996) and Saputro et al. (2020), relying on the foundational work of
Bandura (1986), suggested that what teachers believe about their science capacities
and abilities will influence if and how they successfully engage learners in science
learning. Therefore, aiding preservice teachers in developing their science knowledge and ability, their understanding and beliefs about science, and their science
teaching and learning capabilities is important. Developing positive teacher beliefs
about science and science teaching and learning is desirable. Acting on such recommendations left me thinking about what mix of science knowledge and process should
be involved in the course, and how to engage preservice teachers in deeper learning
relating to science or science teaching.

6.3.2   Science Inquiry

CHO et al. (2011) suggested authentic science inquiry with its problem-solving
experiences can be helpful in addressing epistemological beliefs and understanding
of the nature of science. Such understanding is desirable, if not crucial, for teachers
of science. Silverstein et al. (2009) found that students of inservice teachers who had
authentic experiences of science inquiry had measurable academic improvements as
indicated by scores in later state regency exams. The need for science inquiry also
relates to Windschitl's (2009) call for a radical transformation in science teacher
preparation to involve discourse-intensive collaboration among learners that involves
complex problem-solving or inquiry experiences, learning how to learn, as well as
knowledge development in a subject.

Other science education researchers (e.g., Bell et al., 2003; Bencze & Hodson,
1999; Bergman & Morphew, 2015; Duschl, 2008) have suggested the necessity of
science inquiry experiences for K–12 learners, which again implies that their teachers
should be familiar with such inquiry. Windschitl (2003) was perhaps most direct in

arguing the need for independent authentic science investigations as part of preservice teacher education. These experiences would involve:

- aiding students in engaging in science beyond a confirmatory experience,
- involving preservice teachers with more authentic experiences of science practice,
- supporting preservice teacher autonomy and agency as learners and investigators,
- aiding preservice teachers in developing and investigating an original and unique research question, and
- collaborating with others as part of a science community.

Given such recommendations, I keenly felt the challenge to offer some form of science inquiry to aid preservice teachers in their science learning and teaching.

6.3.3 Nature of Science

Interwoven with the challenge of how to craft inquiry experiences for preservice teachers was my desire to help them develop their understanding concerning the nature of science which Mesci (2020) noted is a key feature of science literacy. Science teachers often have not experienced authentic science inquiry where surface knowledge (Almarode et al., 2018) is deepened and applied to new situations to understand complex relationships and solve problems. Windschitl (2009) noted that preservice teachers are rarely exposed to ideas about science as a discipline at the undergraduate level and do not often participate in discussions of how new knowledge is evaluated. This lack of exposure has been noted before where experience of science is limited to confirmatory laboratory experiences in secondary schools. Learners, including preservice teachers, need to not only investigate the general processes of science but also engage in epistemological conversations about "how we know what we know and why we believe it" (Duschl, 2008, p. 269). This last point directly addresses the need to develop data literacy among preservice teachers.

6.3.4 Provincial Science Curriculum

Alongside considerations of self-efficacy, inquiry, and the nature of science was the likelihood that many of the preservice teachers taking the course would eventually teach the Saskatchewan K–9 Science Curriculum (SME, 2009; Fig. 6.1). The aims, goals, and entry points are typical of other science curricula across Canada. The broad aim of such curricula for K–9 students is to develop scientific literacy. In Saskatchewan curricula, consideration of both Euro-Canadian and Indigenous ways of knowing nature (Aikenhead, 2001) and engagement with Indigenous ways of knowing (Aikenhead & Ogawa, 2007) are important to discussions concerning how science knowledge emerges. Goals specific to understanding the nature of science and science, technology, society, and the environment (STSE) interrelationships;

constructing scientific knowledge; developing scientific and technological skills; and developing attitudes that support scientific habits of mind are included. Such goals and aims suggest that preservice teachers need opportunities to interact with these areas if they are to productively engage their future students.

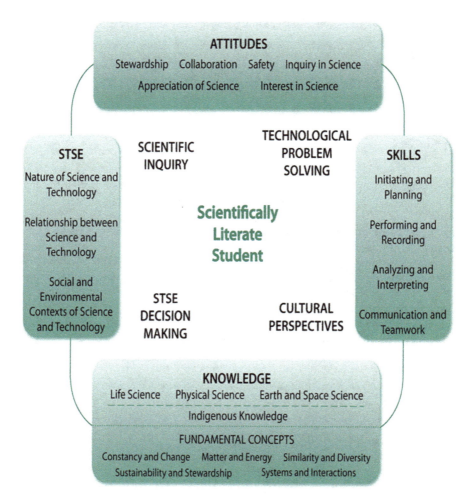

Fig. 6.1 Aims, goals, and entry points for K–12 scientific literacy (SME, 2009, p. 11) (Copyright 2009 by the Saskatchewan Ministry of Education)

6.4 Establishing Learning Outcomes

The previous section addressed factors prompting the development of a new course and revealed motivations for what the course might entail. These factors and motivations positioned my thinking concerning course design. As the team considered how to craft the learning outcomes, the need for preservice teachers to engage in the processes of scientific inquiry became increasingly clear. Our intent was that this engagement would alleviate preservice teachers' possible hesitancy and anxiety concerning science teaching and learning.

We agreed that key to productive inquiry was learner autonomy concerning what might be investigated, which might enhance preservice teachers' engagement and confidence while making science more inviting. We thought engagement might be achieved by focusing on how science exists in daily life, relating directly to learner interests. This approach could support a focus on STSE given intersecting and pertinent topics, for example, climate change, water and food security, human health, energy, and sustainability. This approach also presented opportunities for examining fundamental science knowledge involved in life science, physical science, and earth and space science. Providing opportunities for preservice teachers to develop their science content and process knowledge meshed with our identified need to aid in developing their data literacy.

Eventually, discussions of motivations and intentions led to the establishment of the following outcomes for the course. Preservice teachers would:

1. Investigate a self-selected science topic to develop content and process knowledge relating to the Saskatchewan Grades K–9 science curricula and contemporary STSE topics.
2. Engage in research that included developing an original research question, investigating, and presenting findings.
3. Demonstrate an understanding of the epistemology of science with attention to Western (Eurocentric) and Indigenous ways of knowing.
4. Consider how to critically evaluate scientific knowledge (e.g., scholarly writings, news media).

Within these broad outcomes was the goal of addressing preservice teacher anxiety and concern regarding their ability to teach science. Other goals included developing their confidence as science educators, providing experiences to participate in science inquiry and instructional strategies, emphasizing curricular connections, and developing data literacy.

6.4.1 Other Planning Considerations

With these outcomes in mind, I continued to have questions concerning the nature of the course, specifically issues relating to enrollment, the course as a science or science methods experience, and instructional and assessment approaches. My first concern involved enrollment because this course would enroll 75 first-year preservice teachers, most of whom would not have completed a university science course. Science methods courses at the University of Saskatchewan typically enroll 35 third-year students who have completed some science courses. The larger numbers in this new course seemed at cross purposes with modelling instructional approaches involving a focus on inquiry and hands-on practical activities. The larger enrolment also figured into my concerns about available learning spaces and materials.

My second concern was that, while science and science methods courses are not mutually exclusive, a lack of clarity existed concerning what degree of focus should be on science content knowledge and to what degree this course was to be a science teaching methods course. The decision on course specifics resided mostly with me, as the intended instructor for the course. I was left to interpret the learning outcomes amidst this lack of clarity, which caused some anxiety about how best to proceed. My uncertainty about the nature of the course inevitably affected other aspects of planning, such as instructional approach, content to be covered, assignments, and assessment.

A third concern included selecting instructional approaches that would be productive for preservice teachers in terms of their science learning and eventual science teaching as well as what instructional approaches and experiences would develop their science literacy and science engagement. Elementary-level preservice teachers often describe themselves as lacking ability or affinity for science. I had questions about how to employ inquiry approaches that were engaging and would support the stated learning outcomes beyond a surface science knowledge (Almarode et al., 2018), and I wanted to provide pedagogical experiences where I modeled good science teaching.

My fourth concern involved the interplay between (a) engaging participants in developing science content knowledge and experiencing science inquiry and (b) considerations of pedagogy relating to science teaching and learning, which prompted questions regarding assessment. If there was to be a strong focus on engaging students through inquiry as well as developing science content knowledge, I wondered how best to structure assessments within the learning experiences; for example, the nature and quantity of individual and group assessments and the degree of self, instructor, or peer assessment that should exist. What should I expect in relation to the quality of work of mostly first-year students? How should I assess student effort focused on science process as well as science content knowledge?

6.5   Course Activities

Taking the learning outcomes and the additional considerations into account, I wanted to provide a learning experience where preservice teachers would be challenged in developing their science content and process knowledge through inquiry activities that could be related to the provincial curriculum. I decided on four main teaching activities: a jigsaw activity, problem-solving and laboratory activities, invited speakers, and the University of Saskatchewan's First Year Research Experience program.

6.5.1 Modified Jigsaw Activity

The first activity was a modified jigsaw on essential science concepts. A jigsaw involves a collection of topics that are developed by students before they come together to make a more complete idea (Aronson, 1978). This activity allows individuals or small groups to become responsible for a subcategory concerning a larger topic where they develop and share their knowledge; in this case, a question focused on an essential science concept.

Jigsaw activities can have a positive influence on preservice teachers' learning (Perihan & Tarim, 2007) and can present an opportunity to work collaboratively to develop science content knowledge. The modified jigsaw activity occurred early in the course and allowed the students opportunities to form relationships, experience agency in topic selection, and build individual and group knowledge. Small groups (2–4) investigated a science theory, concept, or phenomenon chosen from a collection of 100 questions split between physical, life, and earth and space science topics related to the provincial science curriculum. Sample questions included: *How does a microwave oven work? What is a vaccination and how does it protect you from a disease? Is there life on Mars?*

Students were encouraged to utilize a variety of media and artefacts they judged helpful in sharing their work and engaging others (e.g., images, graphs, tables, equations, videos, interactive media, worked examples). The modified jigsaw activity culminated in a mini-teaching session of 20 min where each small group presented to several other groups. The entire class did not encounter all the information through presentations. Assessment of this work was split between group and individual reporting and constituted 20% of the final grade.

Student course evaluations revealed that the modified jigsaw activity appeared to be well received and offered opportunities to develop their science knowledge; the opportunity to teach others was mentioned frequently. Students appeared to appreciate the agency they experienced in choosing topics and group members.

6.5.2 Problem-Solving and Laboratory Activities

The second activity involved small group problem-solving and laboratory investigations. Science practice is more than acquisition of book knowledge and involves field and laboratory work where, for example, fundamental experiences of identifying and manipulating variables to investigate a question, concept, or phenomenon can occur.

Engaging with verbal explanations and narratives that involve representations of information are part of this practice.

The problem-solving activities involved students working in groups of 5–6. Two graduate student teaching assistants with backgrounds in science and engineering helped prepare and conduct sessions. Data literacy activities, such as a marble rolling activity (Bowen & Bartley, 2014) or the stabilization wedges game about carbon mitigation (Hotinski, 2015), allowed students to investigate a phenomenon in a manner that yielded opportunities for discussing variables, reliability, validity, and representations of data. Such activities involved collaborative learning and offered me opportunities to discuss science literacy.

The main laboratory work focused on science practices using Vernier LabQuest2 (https://www.vernier.com/product/labquest-2/) and sensor kits. I selected or modified four laboratory investigations from Vernier's materials (i.e., magnetic fields, household acids, conductivity, frictional force) and an additional investigation involving examination of human cheek cells. Students recorded and analyzed the data they generated (e.g., create graphs, perform calculations, find trends). This work provided opportunities to ask meaningful questions about experimental design, scientific phenomena, and use in everyday life. The primary aim of the laboratory investigations was to develop the language, terms, skills, and practices of science as well as to enhance data literacy.

The entire class ($N = 75$–80) was split into two cohorts. During a 3-h class, the two cohorts would switch between laboratory investigations and other activities such as crafting a final laboratory report an individual summary where students commented on their science content and practice development and reflected on the pedagogy they experienced. This report was worth 20% of the final grade.

Teaching assistants reported that students were typically engaged when working through the problem-solving and laboratory activities. End-of-course evaluations revealed mostly positive experiences. To what extent these activities developed science knowledge is less clear as there was not a formal evaluation of content knowledge but a focus on practices of science. The two cohorts were large, which affected access to equipment and involvement in discussion. Also, occasionally there was not enough time for the activities, which affected the quality of the learning experiences. Despite these limitations, the preservice teachers did enter discussions concerning concepts of reliability, accuracy, and precision, which are key ideas in science and in data literacy. Their reflective writing indicated they gained insights into basic science processes and the collection, explanation, and representation of data. The small-group problem-solving and laboratory activities did provide an opportunity to investigate how we come to know what we know in science and to what degree we can rely on this knowledge.

6.5.3 Invited Speakers

The third instructional activity involved invited speakers. During the course, four scientists from the University of Saskatchewan were invited to share their personal story of being a scientist. The intent of including speakers was to engage preservice teachers in thinking about who does science and how science occurs. These speakers came from a variety of backgrounds, genders, and ethnicities. They discussed their research, successes and failures in their academic pursuits, how they persevered, and what their future research might include; they presented data and visual representations from their work and invited discussion about their lives, academic work, and science practices.

The first speaker discussed working with caribou management and Indigenous traditional ecological knowledge. The second speaker talked about food security and asexual seed formation. The third speaker presented their work on walking, balance, and gait analysis. The fourth speaker focused on knee osteoarthritis and computed tomography. These interactions provided an opportunity for preservice teachers to gain insight into current and pertinent science topics and a variety of research approaches. Each speaker shared interesting and engaging stories concerning their life path that eventually led to becoming an expert in their field of study. They also addressed the culture of science research and demonstrated that science is for everyone. End-of-course evaluations contained few comments concerning the invited speakers. However, from my observations during class, students engaged in extended discussions concerning the personal challenges that the scientists face in pursuing their lives and academic work. Students were attentive with few distracted by mobile devices.

6.5.4 First Year Research Experience (FYRE)

The fourth and major activity in the course was the FYRE—a campus-wide initiative for first-year students that fosters active learning through hands-on research. FYRE is centred on the Research Cycle of Question, Investigate, and Share (University of Saskatchewan, 2019). FYRE involved preservice teachers formulating research questions, conducting research, making presentations, and producing a final report. Ideally, students would experience the FYRE Research Cycle as an authentic science inquiry experience involving practical science research (Walker & Molnar, 2013), which would impact their understanding of how we come to know what we know in science.

To formulate their research questions, groups of four to five students considered large topic areas drawn from the provincial science curriculum, such as ecology, sustainability, and industry; climate change; evolution, genetics, and health; food, water, and energy security; and structure and culture. These large topics were then narrowed to local issues, concerns, and contexts that were typically STSE in nature, such as food insecurity in Canadian Northern First Nation Communities, organic farming in Saskatchewan, and impact of post-colonial anthropogenic factors on the South Saskatchewan River. In their investigations, students were encouraged to obtain relevant empirical data from available databases, and represent that data in suitable formats for comparison, evaluation, and re-interpretation. To share their research, preservice teachers created PowerPoint presentations or research posters that were shared with the entire class. Marks for the FYRE inquiry contributed 50% toward the final grade for the course.

A panel consisting of me, the research coach, and the two teaching assistants assessed the FYRE presentations following a rubric. Groups demonstrated varying degrees of sophistication. Despite ongoing feedback and guidance from panel members, many groups reported existing information with limited synthesis or insights. Other groups brought together information in a manner that demonstrated deeper understanding with some synthesis. In several instances, groups synthesized pertinent information from primary sources and databases to develop plausible conclusions. Feedback in FYRE surveys indicated they experienced agency as learners and appreciated the ability to investigate topics of interest. They commented on how FYRE activities developed their confidence in working with others, developing their research skills, and making public presentations.

6.6   Meeting the Learning Outcomes

Responses to the modified jigsaw, group problem-solving and laboratory activities, invited speakers, and the FYRE research project alleviated some of my concerns about modeling good instructional practice and encouraging student engagement. Evidence from student course evaluations, FYRE project surveys, my assessment of student work, and feedback from the teaching assistants and research coach concerning their observations of student engagement suggested that the learning outcomes were realized through the course activities.

All activities and assessments were intended to provide preservice teachers with opportunities to meet the learning outcomes for the course. They had multiple occasions to develop science content and process knowledge in areas related to the provincial curriculum. They followed the FYRE Research Cycle to investigate their own questions and present findings. They gained epistemological insights into the nature of science and Indigenous traditional ecological knowledge. They participated in activities for developing data literacy where they generated and evaluated science knowledge.

6.6.1 *Developing Science Content and Process Knowledge*

Preservice teachers had multiple opportunities to develop science content and process knowledge throughout the course. The modified jigsaw activities focused on the content knowledge related to the science questions that the small groups investigated. My effort to encourage student autonomy by allowing a range of topics worked against their developing similar and sufficient content knowledge. Development of content knowledge tended to be incidental in laboratory and problem-solving activities because they focused primarily on science processes and practices. The invited speakers discussed some science content knowledge and provided insights into science processes, but they focused on the nature of science and how scientific knowledge emerges. The FYRE project also provided opportunities for building science content knowledge, again limited to the research topics selected by the small groups.

My sense is that preservice teachers should be assessed on their science content and process knowledge. However, I did not assess these aspects outside of their reflections upon their own learning. This metacognitive work provided limited opportunities to check science knowledge, but that was not the focus of the assignments. I still have concerns about how best to address this outcome, which is related to my ambiguity related to the overall intent of the course.

6.6.2 *Engaging in Research*

The FYRE project offered opportunities for preservice teachers to engage in research, from developing a question through to presenting findings. The project was intended to make the learning experience more "genuine and pertinent ... more like 'real' science, as practiced by scientists" (Hart, 2002, p. 1240). While all activities undertaken in the course might be considered to involve inquiry or research, the FYRE project was central to achieving this outcome that the preservice teachers met with varying success. Despite my willingness to focus on authentic research as a key feature in the course, and although I sought opportunities for the students to be involved in hands-on research, most groups' FYRE projects resembled reports on existing information rather than self-determined investigations.

The transition from reporting on existing knowledge to applying such knowledge to their research often seemed difficult, which left me to wonder about the approach I had crafted. I even questioned my intent of having students investigate existing data and information to develop new understandings. A more structured process is likely needed, where students are limited in their choice of FYRE topics and have direct encouragement to understand what constitutes an authentic science investigation. However, as evidenced by the final FYRE presentations as well as positive course feedback, this activity successfully allowed students to work together to craft a research question, investigate and make sense of pertinent information, and share results with their peers.

6.6.3 Understanding Epistemologies of Science

The activities in the course provided opportunities for experiencing, discussing, and reflecting on how Eurocentric science knowledge emerges, is structured, and justified. However, despite being specifically identified in the learning outcomes, attempts to engage preservice teachers in discussions of Indigenous traditional ecological knowledge were less successful. Educators involved with Indigenous perspectives and knowledge within Western academic structures, process and goals, and science education specifically know many of the challenges that exist (Aikenhead, 2001; Aikenhead & Ogawa, 2007; Molnar & Jessen Williamson, 2015). Engaging preservice teachers with Indigenous perspective and knowledge within the course, while avoiding tokenism and superficiality, was affected by factors such as time and place restrictions, number of students, protocol challenges, and gaining access to knowledge holders from Indigenous communities.

Opportunities to explore Indigenous perspectives included an invited speaker who discussed their research with Indigenous knowledge holders and scientists working together to investigate climate change and woodland caribou. Another opportunity involved encouraging students to include Indigenous contexts in their FYRE investigations. A few groups took on this work, for example, focusing on food and land security issues of Indigenous communities. Involving students in meaningful investigations of Indigenous epistemologies remains a challenge that is not easily overcome despite a growing number of resources concerning Indigenous and Western understandings (e.g., Kimmerer, 2013; Michell et al., 2008; Weatherford, 2010). Further refinement of course activities to provide clearer direction and more focused opportunities concerning Indigenous ways of knowing is needed in future iterations of this course.

6.6.4 Critically Evaluating Science Knowledge

Opportunities to critically evaluate scientific knowledge included experiences that were intended to develop data literacy. Preservice teachers worked with basic statistics as they completed activities drawn from Bowen and Bartley (2014). They investigated phenomena, generated tables and graphs, interpreted results, and participated in discussions concerning reliability, accuracy, and precision. Preservice teachers were instructed how to utilize popular science websites as a first step towards deeper investigations of primary research articles and media resources that could be accessed through the university library. However, the critical evaluation of the publicly available datasets used with the FYRE project proved challenging because the datasets were often large and did not relate sufficiently to the students' topics. Future iterations of the course need to include ways to better utilize databases.

6.7 Concluding Thoughts

The concerns and challenges that I experienced regarding both the practical and intellectual aspects of this course have lessened after three iterations of teaching it. The central tension—Is it a science or a science methods course?—continues to underlie all other concerns, affecting both my thinking and innate sense about the course. This tension has been alleviated to some extent because my observations of student work and course feedback indicate that these preservice teachers were productive in working with science information through course activities. Comments were generally positive regarding the modified jigsaw, problem-solving and laboratory, and FYRE activities. Opportunities for preservice teachers to critically examine science information, as well as to consider the probabilistic nature of science knowledge, occurred across activities. They engaged in inquiry, experienced a degree of autonomy as learners, participated in collaborative efforts, and discussed how science knowledge comes to be. These experiences provided them with insights into the collaborative nature of science and examples of how they might engage future learners in their classroom.

The course outcome that requires the most pressing rethinking is an understanding of the epistemology of science with attention to Western/Eurocentric and Indigenous ways of knowing. What can be said is: these preservice teachers had fewer opportunities to consider Indigenous ways of knowing, in contrast to Eurocentric perspectives and practices, which were often made explicit and which inherently existed in instructional activities. The course—*Is this a course about science?*—will continue to challenge my thinking and my ability to engage preservice teachers as I navigate the shifting balance between a science or science methods course. I hope what I have shared here can inform other science educators' considerations of their efforts to educate preservice teachers about science and science teaching and learning.

References

Aikenhead, G. S. (2001). Integrating western and aboriginal sciences: Cross-cultural science teaching. *Research in Science Education, 31,* 337–355. https://doi.org/10.1023/A:1013151709605

Aikenhead, G. S., & Ogawa, M. (2007). Indigenous knowledge and science revisited. *Cultural Studies of Science Education, 2*(3), 539–620. https://doi.org/10.1007/s11422-007-9067-8

Almarode, J., Fisher, D., Frey, N., & Hattie, J. (2018). *Visible learning for science, grades K-12: What works best to optimize student learning.* Corwin.

Aronson, E. (1978). *The jigsaw classroom.* Sage.

Bandura, A. (1986). *Social foundations of thought and action: A social cognitive theory.* Prentice-Hall.

Bell, R. L., Blair, L. M., Crawford, B. A., & Lederman, N. G. (2003). Just do it? Impact of a science apprenticeship program on high school students' understandings of the nature of science and scientific inquiry. *Journal of Research in Science Teaching, 40*(5), 487–509. https://doi.org/10.1002/tea.10086

Bencze, L., & Hodson, D. (1999). Changing practice by changing practice: Toward more authentic science and science curriculum development. *Journal of Research in Science Teaching, 36*(5), 521–539. https://doi.org/10.1002/(SICI)1098-2736(199905)36:5%3c521::AID-TEA2%3e3.0.CO;2-6

Bergman, D. J., & Morphew, J. (2015). Effects of a science content course on elementary preservice teachers' self-efficacy of teaching science. *Journal of College Science Teaching, 44*(3), 73–81. http://www.jstor.org/stable/43631942

Bowen, G. M., & Bartley, A. (2014). *The basics of data literacy: Helping your students (and you) makes sense of data.* NSTA Press.

Cho, M-H., Lankford, D. M., & Wescott, D. J. (2011). Exploring the relationships among epistemological beliefs, nature of science, and conceptual change in the learning of evolutionary theory. *Evolution: Education and Outreach, 4*(2), 313–322. https://doi.org/10.1007/s12052-011-0324-7

Duschl, R. (2008). Science education in three-part harmony: Balancing conceptual, epistemic, and social learning goals. *Review of Research in Education, 32*(1), 268–291. https://doi.org/10.3102/0091732X07309371

Hart, P. (2002). Environment in the science curriculum: The politics of change in the Pan- Canadian science curriculum development process. *International Journal of Science Education, 24*(11), 1239–1254. https://doi.org/10.1080/09500690210137728

Hotinski, R. (2015, March 25). *Stabilization wedges: A concept & game.* Princeton University. http://cmi.princeton.edu/wedges/

Jegede, O. J., & Aikenhead, G. S. (1990). Transcending cultural borders: Implications for science teaching. *Research in Science and Technology Education, 17*(1), 45–66. https://doi.org/10.1080/0263514990170104

Kimmerer, R. W. (2013). *Braiding sweetgrass: Indigenous wisdom, scientific knowledge and the teachings of plants.* Milkweed Editions.

McKinley, E. (1996). Towards an Indigenous science curriculum. *Research in Science Education, 26*(2), 155–167. https://doi.org/10.1007/BF02356429

Mesci, G. (2020). Difficult topics in the nature of science: An alternative explicit/reflective program for pre-service science teachers. *Issues in Educational Research, 30*(4), 1355–1374.

Michell, H., Vizina, Y., Augusta, C., & Sawyer, J. (2008). *Learning Indigenous science from place.* University of Saskatchewan.

Molnar, T., & Jessen Williamson, K. (2015). Crossing many boundaries in creating allies: Personal encounters to unfolding science to privilege indigenous knowledge. sihtoskâtowin: kanakiskamihk ôma kâ-taswekinamihk science ekwa mîna kihci-iyiniw-kiskihtamowin. *Education Matters, 3*(1). https://journalhosting.ucalgary.ca/index.php/em/article/view/62962/pdf

Pajares, F. M. (1992). Teachers' beliefs and educational research: Cleaning up a messy construct. *Review of Educational Research, 62*(3), 307–332. https://doi.org/10.3102/00346543062003307

Pajares, F. M. (1996). Self-efficacy beliefs in academic settings. *Review of Educational Research, 66*(4), 543–578. https://doi.org/10.3102/00346543066004543

Perihan, D. A., & Tarim, K. (2007). The effectiveness of Jigsaw II on prospective elementary school teachers. *Asia-Pacific Journal of Teacher Education, 35*(2), 129–141. https://doi.org/10.1080/13598660701268551

Saputro, A. D., Atun, S., Wilujeng, I., Ariyanto, A., & Arifin, S. (2020). Enhancing pre-service elementary teachers' self-efficacy and critical thinking using problem-based learning. *European Journal of Educational Research, 9*(2), 765–773. https://doi.org/10.12973/eu-jer.9.2.765

Saskatchewan Ministry of Education. (2009). *Science 8.* Regina, SK: Author. https://www.edonline.sk.ca/bbcswebdav/library/curricula/English/Science/science_8_2009.pdf

Silverstein, S. C., Dubner, J., Miller, J., Glied, S., & Loike, J. D. (2009). Teachers' participation in research programs improves their students' achievement in science. *Science, 326*(5951), 440–442. https://doi.org/10.1126/science.1177344

Snively, G. J., & Corsiglia, J. (2001). Discovering indigenous science: Implications for science education. *Science Education, 85*(1), 5–34. https://doi.org/10.1002/1098-237X(200101)85:1%3c6::AID-SCE3%3e3.0.CO;2-R

Tippett, C. D. & Milford, T. M. (Eds.). (2019). *Science education in Canada.* Springer. https://doi. org/10.1007/978-3-030-06191-3

University of Saskatchewan. (2019). *FYRE: First year research experience.* https://vpresearch. usask.ca/students/undergraduate/documents/FYRE-Faculty-Package.revised.MM.July2019.pdf

Valls-Bautista, S.-L. A., & Casanoves, M. (2021). Pre-service teachers' acquisition of scientific knowledge and scientific skills through inquiry-based laboratory activity. *Higher Education, Skills and Work-Based Learning, 11*(5), 1160–1179. https://doi.org/10.1108/HESWBL-07-2020-0161

Walker, T., & Molnar, T. (2013). Can experiences of authentic scientific inquiry result in transformational learning? *Journal of Transformative Education, 11*(4), 229–245. https://doi.org/10. 1177/1541344614538522

Weatherford, J. (2010). *Indian givers: How the Indians of the Americas transformed the world.* Crown.

Windschitl, M. (2003). Inquiry projects in science teacher education: What can investigative experiences reveal about teacher thinking and eventual classroom practice? *Science Education, 87*(1), 112–143. https://doi.org/10.1002/sce.10044

Windschitl, M. (2009, February 5–6). *Cultivating 21st century skills in science learners: How systems of teacher preparation and professional development will have to evolve.* National Academies of Science Workshop on 21st Century Skills. University of Washington. https://sites. nationalacademies.org/cs/groups/dbassesite/documents/webpage/dbasse_072614.pdf

Tim Molnar BSc Advanced (University of Saskatchewan), BEd (University of Saskatchewan), MEd (University of Regina), PhD (University of Victoria), is an assistant professor in the department of Curriculum and Instruction at the University of Saskatchewan. For many years he taught secondary school science in Treaty Four territory. Currently he teaches science methods for elementary and secondary school preservice teachers and graduate courses in curriculum theory. His research involves investigating student experience in undergraduate science and science education coursework, examining understandings of responsibility in intercultural contexts in relation to science, and more recently investigating student and community engagement in climate monitoring projects undertaken by Indigenous communities.

Chapter 7
Professional Learning Using a Blended-Learning Approach with Elementary Teachers Who Teach Science: An Exploration of Processes and Outcomes

Xavier Fazio◉ and Kamini Jaipal-Jamani

7.1 Introduction

Professional development programs (PDPs) are purposeful learning experiences that require tremendous resources from schools and districts. One important priority for science education is to improve the pedagogical knowledge and skills of practicing science teachers through PDPs. An evolving understanding of how best to teach science calls for a significant transition in the way science is currently taught in classrooms and will require many science teachers to innovate how they currently teach science (Darling-Hammond et al., 2017; Luft & Hewson, 2014; National Academies of Sciences, Engineering, & Medicine [NASEM], 2015). One way for science teachers to develop their professional practice is through a twenty-first century learning approach using a blended (i.e., online and face-to-face) program. While not new (e.g., Boitshwarelo, 2009; Hotze et al., 2020; Owston et al., 2008; Psillos, 2017), this model has never been researched as part of a large-scale PDP implemented by a teaching association in Canada. Blended PDP has been initiated by other teacher professional associations (e.g., National Science Teaching Association, 2020), but results from such studies may not be fully applicable across contexts. Conducting research that is specific to teaching communities and diverse populations is also important and necessary for Canadian educators to improve classroom teaching practices (Campbell, 2017).

X. Fazio (✉) · K. Jaipal-Jamani
Faculty of Education, Department of Educational Studies, Brock University,
St. Catharines, ON, Canada
e-mail: xfazio@brocku.ca

K. Jaipal-Jamani
e-mail: kjaipaljamani@brocku.ca

Our evaluative research focused on a PDP, herein referred to as 'Science Teaching Innovation' (STI); its three main elements were implemented in chronological order: face-to-face presentations and workshops, collaboration and coaching via an online learning platform, and knowledge mobilization through resource sharing of grade-specific curriculum artifacts. STI's program focus was consistent with science curriculum reform initiatives including Canadian and USA standards documents (e.g., British Columbia Ministry of Education, 2019; Luft & Hewson, 2014; National Research Council, 2013; Ontario Ministry of Education, 2008, 2022), which foreground student sense-making and the emergence of science concepts from classroom practices and events that support knowledge development with students in schools (Sandoval, 2015; Thompson et al., 2016). This study addressed the importance of teacher professional learning in Canada and supports recommendations regarding the readiness of Canada's teachers and students to meet future skill requirements in science, technology, engineering, and mathematics (STEM; Council of Canadian Academies, 2015).

While the STI program involved K–12 science teachers, the focus of this chapter is to highlight research results evaluating its impact on elementary science teachers. Based on analyses of these results, recommendations will be provided regarding design principles for blended-learning PDPs appropriate for PK–12 educators. Furthermore, this study addresses the importance of future research regarding science teacher professional learning in Canada.

7.2 Background and Conceptual Framework

The research on teacher PDPs is extensive; over the last two decades, researchers have reached a consensus on key features of highly effective, in-person programs. Key PDP features are situated in classroom practice, focused on student learning, embedded within professional learning communities, and sustained over time (Bybee & Loucks-Horsley, 2000; Guskey & Yoon, 2009; Hill et al., 2013). Other features of effective PDPs include active learning workshops based on research-based practices, teacher study groups, and discipline-specific content and pedagogy (Darling-Hammond et al., 2017; NASEM, 2015; Penuel et al., 2007). Furthermore, Windschitl (2009) maintained that the above features—together with the collaborative participation of science teachers in the same school, grade, or subject areas—support students' development of twenty-first century competencies.

Despite a consensus in the literature on PDPs, there is limited evidence of the specific characteristics of PDPs that use a blended-learning approach (Community for Advancing Discovery Research in Education [CADRE], 2017; NASEM, 2015; Wayne et al., 2008). Indeed, little research is available on how to develop science teaching and technology integration effectively within a blended-learning approach (NASEM, 2015). The limited research findings do suggest that there is promise in using a blended-learning strategy for PDP design in science and mathematics (Hodges et al., 2013; Owston et al., 2008; Sinclair & Owston, 2006). Research also suggests that PDPs are enhanced by a network of communication between facilitators and teachers, along with face-to-face connections amongst the teachers themselves (Luft & Hewson, 2014; NASEM, 2015). For this chapter, we will briefly describe the practices promoted through STI's blended framework: scientific inquiry and technological design, technology-mediated pedagogies, critical thinking, and other twenty-first century skills.

7.2.1 Scientific Inquiry and Technological Design

Scientific inquiry and technological design pedagogies are generally defined as approaches that provide students with opportunities to engage and understand scientific and engineering practices regarding natural or human-designed phenomena (e.g., Alberta Education, 2014; NRC, 2013). These approaches are typically experiential in nature and can be organized into recognizable pedagogical models such as inquiry-, problem-, design-, and project-based learning approaches. Often the teacher and students work collaboratively with respect to the amount of guidance, personalization, and performance features required when implemented in science classrooms. Further, these models are oriented toward meaningful understanding of scientific or engineering concepts along with developing expertise in students' planning, performance, and communication competencies for future application in diverse school and community contexts (Organization for Co-operation and Development [OECD], 2017; Peterson et al., 2018).

7.2.2 Technology-Mediated Pedagogies

Technology-mediated pedagogies are broad, learner-centred approaches that use information and communication technologies (ICT). These technologies have become pervasive in our schools and society (People for Education, 2019). In a science classroom, for example, this may include the use of tablets, laptops, desktops, smartphones, projectors, video conferencing apparatus, 3-D printers, robots, and probeware sensors. Connecting these devices to local and cloud-based software for students to individually or to collaboratively access online information and multimedia has become reasonably common practice in schools. Importantly, the use of technology-mediated pedagogies to enhance, rather than distract, students' science learning has become an important goal for science teachers (Krajcik & Mun, 2014). In addition, these technology-mediated pedagogies connect to both didactic and experiential pedagogies. For example, independently learning science content by viewing a lecture on video scaffolded with a series of structured questions is a common online didactic pedagogy; team collaboration on an online discussion platform to engage in problem- or project-based learning is a common experiential pedagogy (Bates, 2018).

7.2.3   Critical Thinking

Critical thinking in education is a popular program goal that involves cognitive processes that pedagogies help to promote. According to Hitchcock (2018), it is more useful to describe characteristics of critical thinking rather than try to come up with a succinct definition. In general, critical thinking has the following features:

- It is a goal-oriented process for the purpose of deciding upon future actions or views.
- The student engaging in the thinking is trying to fulfill standards of accuracy appropriate to the thinking of the discipline in relation to scientific propositions or claims.

Often critical thinking is conflated with Bloom's evaluating thinking category; however, it necessitates metacognitive thinking (Ennis, 2016). Students in science might employ critical thinking processes by inferring, experimenting, observing, hypothesizing, researching, and evaluating scientific phenomena and claims.

7.2.4   Other 21st Century Skills

The United States NRC (2013) has categorized twenty-first century skills into three areas of competency—cognitive (e.g., critical thinking skills), intrapersonal (e.g., metacognition skills), and interpersonal (e.g., communication)—and recommends a deeper learning approach to develop these competencies. Deeper learning or meaningful learning is described as the process through which an individual becomes capable of taking what was learned and applying it to new situations (i.e., transfer) (NRC, 2013). A mutual, reinforcing relationship exists between deeper learning and twenty-first century competencies. The NRC further recommends the use of pedagogical strategies such as modeling, guided inquiry, and technology-mediated activities to engage in deeper learning to develop these skills. The PDP in this study was informed by the need to provide teachers with knowledge of, and practice with, inquiry and engineering design practices along with technology-enhanced strategies to support deeper learning in school science.

7.2.5   STI Professional Development Program Features

An operational model of how PDPs can impact teachers was proposed by Desimone (2009). In this model, teachers participate in programs that in turn increase their knowledge and skills or change their views on science instruction. From this, teachers adapt their instructional practices and subsequently can impact student outcomes. Of importance to these efforts is the sequence in which these events occur (Guskey,

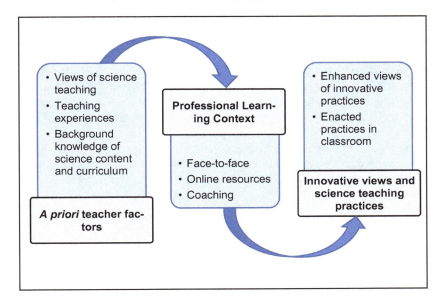

Fig. 7.1   Framework for the STI professional development program

2002). While this model does not take into consideration all the complexities of delivering PDPs, such as school contexts (e.g., Fazio, 2009; Sandholtz & Ringstaff, 2016), the STI program was primarily designed using this model (Fig. 7.1). We agree with Desimone's proclamation that this model is still worthy of careful investigation as it would elevate the quality of PDP studies, particularly in Canada, where there is a scarcity of such research.

Noteworthy for this study was how the STI program promoted innovative science teaching using a blended approach that included face-to-face interactions, online communication, and support though the teaching association's model of online coaching. Nevertheless, blended PDPs can vary considerably; and evidence is still sparse on its impact on science teacher practices (NASEM, 2015; Wilson, 2013). For this study, blended learning was demonstrated as an application of content ideas online with coaches (i.e., facilitators) after face-to-face interactions with teachers at an annual professional conference and prior to their pedagogical sharing at the next annual conference.

7.3   Methodology

The objectives of this research were to (a) evaluate the impact of the PDP on participating science teachers' views and practices and (b) provide recommendations for future professional learning opportunities for inservice science teachers. Thus, the focus of this research project was evaluation research versus basic or applied research. In general, the goal of evaluation research is to make judgments on specific programs and provide recommendations for program providers or external agencies (McMillan, 2004). In this study, our research roles were circumscribed as long-standing and trusted colleagues with the science teaching association delivering the STI program; yet we maintained a professional distance from the PDP's design and implementation. Finally, this study was reviewed and approved by a university ethics review board.

The study utilized a mixed-methods approach employing a concurrent design with qualitative and quantitative data, separate data analysis, and integration of the two data types (Creswell, 2012; Teddlie & Tashakkori, 2009). The research objectives were to:

- Assess changes in the elementary teachers' science educational views (efficacy, instruction) and practices (inquiry, technology-enhanced, critical thinking) based on their participation in the PDP
- Identify strategies and practices enhancing the effectiveness of blended professional development based on the research findings.

Over the two years of data collection (2015–2017), there were 142 elementary teacher participants (Year 1 = 59; Year 2 = 83) with four cohorts of approximately 20–30 participants per year organized in grade groupings: Kindergarten to Grade 3 (primary), Grades 4 to 6 (junior), Grades 7 and 8 (intermediate). The participants were elementary teachers from across the province who took part voluntarily although some participants were selected to attend based on centralized school district decisions. Note that grades up to Grade 8 are typically considered part of the elementary panel in the province. Note that we do not identify the province in order to maintain our anonymity requirements for this study.

7.3.1 Timeline, Data Sources, and Analysis

A timeline of the project and delineation of the research activities undertaken for the study are shown in Fig. 7.2. The study was undertaken in the 2015–2016 and 2016–2017 academic years involving two different groups of science teachers. Data sources for the study included online surveys using SurveyMonkey™ prior to the implementation of the program and after completion of the online phase. In addition, there were written evaluations of workshops at the face-to-face program, analysis of discussions posted on STI's online learning platform, interviews with teacher participants representing all the cohorts noted earlier during the online phase of the PDP, document analysis of teacher-developed classroom resources, post-program surveys, and interviews with select teachers at the sharing sessions. Due to page restrictions for this chapter, we cannot present a full representation from the data corpus. Summaries are provided with respect to the data sources and results. The

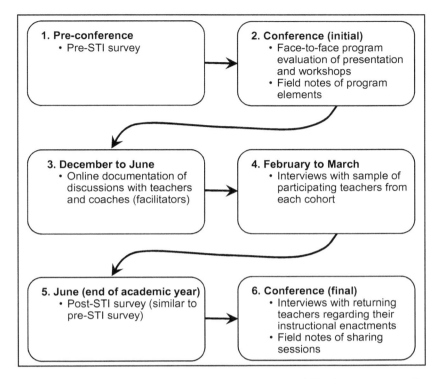

Fig. 7.2 STI program features and summary of data collection activities for a one-year cycle

question themes used for the various data collection activities identified are provided in Table 7.1.

All qualitative data sources (e.g., interviews, online discussions, open-ended questions from pre/post surveys) were transcribed. Text analysis involved coding and categorization of recurring themes using a multistage and iterative procedure guided by the processes of exploring and describing the interpreted qualitative data (Creswell, 2012; Merriam, 2009), facilitated by NVivo™ software.

The quantitative data sources (e.g., pre/post surveys) were analyzed using IBM SPSS Statistics. Quantitative analyses involved descriptive, parametric, and non-parametric statistics appropriate to the data sets produced from the pre/post program surveys. Descriptive statistics were produced for each quantitative-based survey question. This included frequency and percentage distributions for all items on Likert-type scales or rank-order questions. The descriptive statistics included mean, standard deviation, and composite scores where applicable. To address the research objectives, the primary focus of the inferential statistical analyses was on group comparisons before and after the STI program, that is, pre/post program comparisons. Unless noted, the data comparisons were overall comparisons of the entire program (i.e., Years 1 and 2 data combined). Owing to sample-size considerations, comparisons

Table 7.1 Data collection instruments and questions themes

Data collection instruments	Question themes
STI surveys (44 questions) Likert-type Rank ordering Open-ended questions	Demographic Past professional development experiences in science Science Teaching Efficacy Belief Instrument (STEBI- Elementary) Type and frequency of various of forms of instruction and assessment Confidence with and frequency of using scientific inquiry, technological (engineering) design teaching Confidence with and frequency of using technological tools for instruction Confidence with and frequency of teaching critical thinking Instructional resources used Obstacles to teaching science
Face-to-face evaluation of workshops (initial conference)	Relevance of topic to instruction Impact on knowledge of topics (e.g., inquiry, technology enhanced instruction) Quality of workshop Clarity of STI PDP aims and next steps
Interviews with sample teachers (midpoint)	Demographics and professional qualifications Past professional development experiences in science Views of and frequency with scientific inquiry/technology design teaching, technological-enhanced instruction, critical thinking instruction Current views of the STI PDP and progress to date Use of the online platform for professional learning Challenges of STI PDP and new understandings
Interviews (final conference)	Review of instructional activities/unit (enactments) implemented in their classroom Identification of any new learnings based on their instructional enactments Benefits and challenges of STI PDP

among smaller cohorts (e.g., Kindergarten–Grade 3 vs. Grades 4–6 cohorts) were not deemed statistically sufficient.

A limitation of the study was that data were not collected in school settings due to restrictions in our research roles based on the scope of the evaluation project. Other limitations include self-reporting bias in the surveys, and the potential for participant bias based on the convenience sample for the interviews based on researcher selection, which may have been further impacted because of the attrition of participants in some cohorts. Nevertheless, since data were collected across multiple cohorts over two

academic school years, more robust conclusions can be drawn about the effectiveness
of a blended PDP. Additionally, there was high confidence in our capability to address
the research objectives.

7.4   Results

As identified earlier, a concurrent mixed-methods research design was used for this
research study. We highlight our analyses appropriate to the themes presented in
this section. The results provided are exemplars only due to page restrictions. Our
analyses of data were collected from participants in either Years 1 or 2 of the STI
program—different groups of elementary teachers. Our results consist of examples
that includes participants' responses from an evaluation questionnaire of the face-to-
face program sessions, interviews with convenience-selected participants (selected in
a randomized manner and contacted for availability) from various cohorts, interviews
with participants presenting at the "Sharing the Learning Showcase" in Fall 2016 and
2017, responses to open-ended questions on the survey administered to participants
at the end of the PDP, participants' engagement data on the STI online platform,
and pre- and post-STI survey data. The findings below are supported with exemplars
and are organized according to the thematic categories from our analysis: changes
in pedagogical views and practices, teaching self-efficacy changes, active learning,
and motivating and sustaining engagement online.

7.4.1   Changes in Pedagogical Views and Practices

Based on analyses of both quantitative and qualitative data, the STI program had
modest impact on these teachers' views with respect to innovative teaching practices.
More specifically, fine-grained analyses of pre/post program survey data comparisons
(see Fig. 7.3 for results and survey questions) illustrate that these elementary teachers
improved their views significantly with respect to their teaching effectiveness of
inquiry- or design-based science teaching and certain technology-enhanced teaching
to address the science curriculum. What we mean by significance is that there is a
5% probability that the results are due to chance alone. For survey question about the
teachers views (i.e., agreement with) about various innovative instructional practices
(Fig. 7.3), the Mann–Whitney U test statistic was $M = 3.98$ (pre), $M = 4.32$ (post)
$U = 1885, z = -2.325, p = 0.02$ for their views about using inquiry or design
teaching to address the Ontario curriculum, and teaching with technology to address
the Ontario curriculum was $M = 3.78$ (pre), $M = 4.11$ (post) $U = 1887, z = -2.366,$
$p = 0.018$. We consider these as significant increases, $p < 0.05$. However, they did
not improve significantly pre/post program with respect to their views of inquiry and
technology pedagogies to develop students' critical thinking skills or understanding

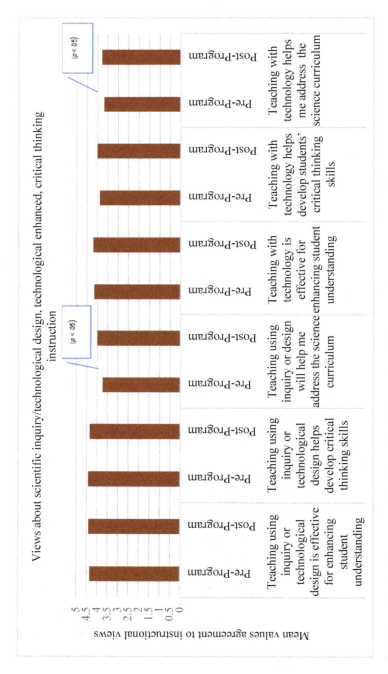

Fig. 7.3 Mean values for elementary teachers' instructional views pre/post STI program about scientific inquiry/design and technology instruction to address curriculum and critical thinking

in science. While there were positive trends in other instructional views pre/post STI program, statistical comparisons found no significance.

Elementary teachers were asked, "How often do you incorporate various inquiry-based instructional strategies into your science instruction?" These strategies included using deductive and inductive approaches, developing students' questions, having students make claims using evidence, and communicating their results. While elementary teachers in both years reported a positive trend in frequency of use of inquiry-based strategies in their respective classrooms, overall, there was no statistically significant differences pre/post STI program for both Years 1 and 2 groups with respect to inquiry elements. However, there was one exception: reported frequency of allowing students to develop their own questions.

It is clear that these elementary teachers' instructional views improved regarding using inquiry and engineering design as well as teaching with technology to address the science curriculum. While there was a positive trend for views of technology-enhanced learning practices in their respective classrooms, there was no significant difference reported in frequency of technology use post-STI program. These findings were partially corroborated with data from interviews where teachers highlighted changes in their innovative instructional views and practices due to the STI program.

> It is a great program [STI] that compelled and motivated me to adapt and evolve my teaching to empower learners with the skills needed for future (Junior teacher, sharing showcase).
>
> It helped me better understand what inquiry is and what it might look like in the science classroom (Intermediate teacher, sharing showcase).
>
> It inspired me to be more hands-on and innovative (Intermediate teacher, sharing showcase).

However, comments after the face-to-face workshops portrayed a mixed picture of the impact of the initial component of the PDP.

> I would have liked to have seen more examples of how to incorporate technology and inquiry into my classroom (Primary teacher, face-to-face evaluation questionnaire).
>
> Lecture style is not valuable. You should emulate what you want students to do! For science, explore and wonder first and then come back and consolidate over the learning together like you would in a classroom. PDP should follow this model (Intermediate teacher, face-to-face evaluation questionnaire).
>
> More hands-on activities and less 'theory' presentations (Junior teacher, face-to-face evaluation questionnaire).

Of interest were findings that elementary teachers demonstrated significant decrease, $M = 2.37$ (pre), $M = 2.09$ (post) $U = 7035, z = -2.998, p$ 0.04, when comparing pre/post program surveys regarding their frequency of use of teacher-directed and surface-level instructional practices (e.g., teacher explaining science concepts, memorizing science vocabulary). This evidence highlights the beginning of a shift in instructional classroom practices (Guskey, 2002). Nevertheless, significant shifts in instructional competencies related to the innovative pedagogical foci (i.e., inquiry, technology enhanced, critical thinking) due to the STI program were not found in the survey data.

7.4.2 Teaching Self-Efficacy Changes

The online survey had one set of Likert-type statements that measured participants' self-efficacy based on the Science Teacher Efficacy Belief Instrument (STEBI;
Riggs & Enochs, 1990). A subset of these statements yields scores for the Personal
Science Teaching Efficacy (PSTE) subscale, which reflects science teachers' confidence in their ability to teach science. The other subset yields scores for the Science
Teaching Outcome Expectancy (OE) subscale, which reflects science teachers'
beliefs that student learning can be influenced by effective science teaching. Data
from the STEBI subset of survey questions uncovered that Year 1 elementary
teachers' science teaching efficacy scores increased significantly, paired samples
t-test $M = 3.8692$ (pre), $M = 4.2708$ (post), $p < 0.01$. However, the Year 2 elementary teachers efficacy scores did not change significantly even though there were
efforts made by the teaching association to improve the STI program in terms of
differentiating the face-to-face component of the PDP for the participating teachers.

7.4.3 Active Learning

Face-to-face sessions provided pragmatic pedagogical strategies for designing
inquiry and engineering design-based curriculum units, which spurred changes in
teaching practices for some teachers. An intermediate science teacher from the
sharing sessions described how ideas gained from workshops at the beginning of
the program influenced their views about innovative science teaching and their
corresponding practices:

> I would not have done this [classroom engineering project] without the STI program. A
> workshop speaker showed us how inquiry and engineering practices looked and how you
> stage it, so you do it step-by-step. Whereas in our schools we have always been told to do
> inquiry and you are kind of left to do your own inquiry … it was an eye opener. [Intermediate
> teacher, key participant interview]

The knowledge initially gained from the face-to-face sessions enabled these
teachers to modify their teaching practices prompted by active learning experiences so that they could in turn engage their students in self-directed inquiry and
engineering-based learning. The evidence of student engagement in inquiry and
design-based learning was showcased at the end of the PDP, where participants shared
the inquiry/design or technology-enhanced student learning activities or projects they
implemented within their respective classrooms. Figure 7.4 shows how a teacher
participant engaged students to solve the problem of creating a launch pad that would
allow rockets to go the furthest distance. The knowledge she gained from the STI
program enabled changes to her teaching practices so that she was able to engage her
students in self-directed inquiry—something she had not done before. As the teacher
explained, "Previously, I felt that I had to direct them because they wouldn't learn if

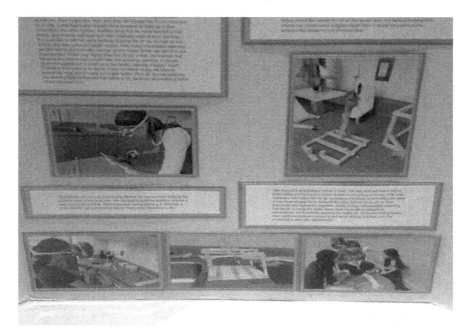

Fig. 7.4 Poster presentation from an elementary teacher showcasing student inquiry and design projects

I didn't." These findings showcase the importance of active learning and its impact on pedagogical change for teachers (Darling-Hammond et al., 2017).

7.4.4   Motivating and Sustaining Engagement Online

Prior to the elementary teachers' participation in STI, they were surveyed about their experiences with professional learning in an online environment. Only one-third said they would consider learning online, and one-half said that they would not consider this form of learning. The online PDP platform was an important and novel professional learning feature for the STI program. Supporting the online activity were designated online mentors who were experienced science teachers. These individuals were responsible for facilitating cohort discussion (i.e., Grades K–3, 4–6, 7–8) and supporting online teacher learning after the face-to-face participation until the end of the academic year. By the end of the program, participation with the STI online activities seemed to improve only marginally, as shared by one participant in an interview.

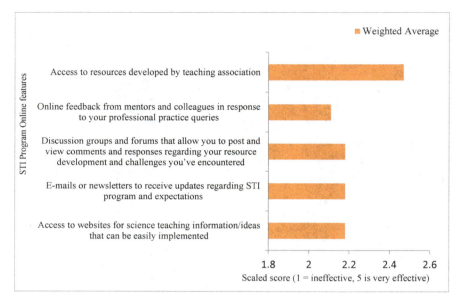

Fig. 7.5 Effectiveness of the STI program online features for improving professional practices

> I've been on it [online platform]. I found it very frustrating to be honest with you. Was it the technical? the structure? I can't put my finger on it. I've been on it and I just did not find it enjoyable, the interface. I think that's it. It's the interface that threw me off. I think if it were simpler then I would use it more. (Junior teacher, key participant interview)

This quote expressed the uncertainty and frustrations that many of the participating teachers experienced with the online features of the STI program. The survey results (Fig. 7.5) further illustrate the challenges inherent in various online features of the program.

7.5 Discussion

The STI professional learning program identified innovative teaching practices relevant to many twenty-first century teaching priorities for science education (i.e., inquiry-based, technology-enhanced, critical thinking). Based on analyses of both quantitative and qualitative data, the teaching association's PDP program had modest impact on elementary teachers' views and associated instructional practices. Elementary teachers participating in the project improved more than their secondary science teacher counterparts (not included in this chapter's review). Specifically, modest changes toward more progressive views of inquiry-based science and technology-enhanced instructional views were evident with these elementary teachers by the end of the STI program. The same cannot be said for teaching critical thinking, where little evidence of changes for this professional learning goal was found in the teachers' views or practices data.

Inferring from the survey data, there were more improvements with elementary teachers in Year 1 than the subsequent Year 2 group, as detected by their teaching efficacy scores. Still, based on the aggregated data collected through the surveys and interviews across both years of the STI program, significant shifts in either group's science teaching efficacy were not found. Corroborating these findings with participant interviews highlights an important research finding from this study: improving general science teaching efficacy may not be a sufficient condition toward adopting innovative science teaching practices.

Another interesting finding is that these elementary teachers demonstrated statistically significant decreases in teacher-directed and surface-level instructional practices (e.g., teacher explaining science concepts, memorizing science vocabulary). In addition, the teachers began to show confidence (as measured through the pre/post surveys) in dealing with obstacles to teaching science (e.g., inadequate materials/equipment). This evidence highlights the beginning of a shift in instructional classroom practices and is an important external marker documenting professional learning (Guskey & Yoon, 2009). Nevertheless, significant shifts in instructional competencies related to the innovative pedagogical foci (i.e., inquiry, technology-enhanced, critical thinking) of the STI program were not found in the data collected.

From an individual case versus aggregated perspective, some individual elementary teachers benefitted from their participation in the STI program and modified their practices. These teachers showcased their innovative pedagogical practices and curriculum development experiences at the sharing sessions at the end of the program. After interviewing some of these teachers, many can be described as motivated and willing to embark on new learning when presented with a professional learning opportunity. This finding aligns with current research on teacher competencies and innovative teaching where professional learning, cooperation, ethical, and communicative competencies are highly correlated with professional learning and the implementation of innovative teaching practices (Zhu & Wang, 2014). Furthermore, some of these motivated individuals were already part of similar professional learning initiatives in their respective school or district prior to their participation in the STI program. These observations speak to the importance of differentiating professional learning for teachers when they voluntarily participate in PDPs (Grierson & Gallagher, 2009).

The opportunity for science teachers to explore and engage in active, in-person learning activities (e.g., doing inquiry or practice using digital technologies) designed for their students was clearly relevant to the participants. Nonetheless, interviews and survey results highlighted that even modest improvements in general science teaching efficacy may not be sufficient for adopting these specific innovative teaching practices. This finding resonates with other PDP research and the impact of teachers' respective school cultural and political contexts (Fazio & Gallagher, 2018; Penuel et al., 2007).

One challenge observed in the STI program was sustaining teacher engagement after the face-to-face components were completed. Indeed, the attrition rate for the teachers who either did not complete the final program elements or participate in the final survey was approximately 30%. A possible explanation could be that when teachers were challenged with implementing innovative teaching practices, the cognitively and emotionally demanding experiences made it more difficult to sustain engagement, especially when the PDP had an online requirement. This explanation has been found with other blended PDPs (e.g., CADRE, 2017; Hodges et al., 2013). Furthermore, technological demands for online learning (e.g., using web-based platform and online communication tools) exacerbated this situation. Indeed, the participants who struggled yet sustained their engagement with the STI program commented in interviews regarding this specific challenge.

> The STI project connected teachers from different schools in the same area but, because there was no release time for us to connect together, it fell apart. So, then I had to make my own group at school [with teachers not part of STI]. (Intermediate teacher, key interview participant)

> I think that if I had a partner at my school the program would have been more successful for me. (Junior teacher, key interview participant)

The literature reports that a structured yet flexible online learning environment provides a better engagement platform than an overly flexible and less-defined experience (Owston et al., 2008), which was evident with the STI program.

Critical to any successful PDP is the transfer and implementation of teachers' learning into their school context. An organizing framework is required to help elementary science teachers with this process. While outside the chapter focus, promising work with respect to using a collaborative inquiry research approach (i.e., collaborative action research or design-based research) is a promising strategy for professional learning that science educators can use to adapt existing practices and adopt innovative teaching approaches aligned with curricular and pedagogical reforms and that are relevant to their school context (Fazio, 2009; Fazio & Gallagher, 2019; Goodnough, 2018; Maeng et al., 2020).

7.6 Implications and Recommendations

Designers of PDPs face many decisions when planning professional learning, yet there may be little evidence from the research corpus to support these various decisions (Hill et al., 2013). Our findings provide specific recommendations and contribute to the research call aimed to identify evidence-based practices from large-scale studies in blended-learning PDPs for K–12 science teachers (Polly & Hannafin, 2010). Furthermore, this study investigated research conjectures for teacher change and instruction, which are necessary to contribute to our understanding of how PDPs work, especially in blended-learning environments (Desimone, 2009; Wayne et al., 2008).

Based on our study, and supported by existing literature, we identified seven specific recommendations for blended PDPs to improve science teacher engagement in the STI project. Some recommendations may be applicable to other professional development projects with study designs similar to that used in the STI project. We recommend the following strategies:

- Provide technological supports to practice with online applications required for collaborating.
- Create mutual accountability structures, such as division of group tasks and timelines amongst participants, to maximize productivity and encourage online collaboration.
- Integrate face-to-face synchronous and asynchronous meetings during the online component.
- Plan online events that use learning objects to focus pedagogical discussions and subject-specific practices.
- Utilize online tasks that encourage reflection surrounding a curriculum product and student work.
- Modify the professional development expectations based on the social-cultural perspectives of the schools and districts where the teachers are working.
- Provide more initial face-to-face workshops that involve hand-on experiences with topics (e.g., engineering design) and technology-enhanced pedagogy that increase the confidence and capabilities of the teachers prior to embarking on online PDP experiences.

The science teaching association providing the PDP is dedicated to improving the teaching of science and providing professional learning opportunities and resources for science educators. To that end, the STI design team embarked on an ambitious journey of professional learning for science teachers. Acknowledging teachers' preservice and inservice science experiences, adapting to school contexts and provincial ministry of education policies, and honouring professional motivations and classroom experiences is difficult to accomplish. Recognizing that teacher learning is a complex endeavour is an important consideration for future professional learning programs. To be effective, adequate time must be provided for science teachers to learn, practice, and implement innovative teaching strategies (Darling-Hammond et al., 2017; Mutch-Jones et al., 2022).

As a professional learning opportunity, the STI program provided a catalyst for motivated science teachers to further their knowledge regarding innovative pedagogies. However, many of the participating science teachers were unable to sustain their engagement with the PDP, particularly after the face-to-face component was completed. While the STI online platform was a novel and potentially useful learning platform for teachers, specific online and other accountability practices (e.g., required online meetings, periodic face-to-face sessions) are essential to effectively use this digital twenty-first century learning approach. Almost certainly, future designs that use a blended-learning approach must take into consideration effective design features (e.g., flexible yet accountable online requirements, sharing of curricular exemplars online, frequent feedback on pedagogical practices) that have worked successfully in other intensive online PDPs (CADRE, 2017; Fernandes et al., 2020; Owston et al., 2008).

Attention to the integrated recommendations from the participants and extant research literature (e.g., NASEM, 2015) is essential for future PDP providers who may wish to develop programs for science teachers. Indeed, some of the successful and unsuccessful design elements have been already tested by many existing programs in Canada and abroad (Campbell, 2017). Substantially less research exists on opportunities for professional science teacher communities in schools, mentoring and coaching, online learning, and science teacher networks (NASEM, 2015). This research project sheds light on the nebulous area of professional learning for science teachers, especially relevant in Canada where very few studies exist. With a shifting educational policy agenda in Canada and abroad, professional learning for science educators must be redesigned to meet science students' needs. We see this study as the beginning of an important trajectory for the science education community. Only through a system of collaboration with practicing science teachers—along with evidence-informed feedback from researchers and PDP providers—will there be professional learning that supports comprehensive adoption of innovative science teaching practices in Canadian schools.

References

Alberta Education. (2014). *Science grades 7–8–9: Program of studies.* https://education.alberta.ca/media/3069389/pos_science_7_9.pdf

Bates, A. T. (2018). *Teaching in a digital age: Guidelines for designing teaching and learning* (2nd ed.). Creative Commons Attribution NonCommercial.

Boitshwarelo, B. (2009). Exploring blended learning for science teacher professional development in an African context. *The International Review of Research in Open & Distributed Learning, 10*(4). https://doi.org/10.19173/irrodl.v10i4.687

British Columbia Ministry of Education. (2019). *BC science curriculum.* https://curriculum.gov.bc.ca/curriculum/science

Bybee, R. W., & Loucks-Horsley, S. (2000). Advancing technology education: The role of professional development. *The Technology Teacher, 60,* 31–34. http://citeseerx.ist.psu.edu/viewdoc/download?doi=10.1.1.680.5344&rep=rep1&type=pdf

Campbell, C. (2017). Developing teachers' professional learning: Canadian evidence and experiences in a world of educational improvement. *Canadian Journal of Education, 40*(2), 1–33. https://journals.sfu.ca/cje/index.php/cje-rce/article/view/2446

Community for Advancing Discovery Research in Education. (2017). *Emerging design principles for online and blended teacher professional development in K-12 STEM education.* Education Development Center.

Council of Canadian Academies. (2015). *Some assembly required: STEM Skills and Canada's economic productivity.* Council of Canadian Academies Expert Panel on STEM Skills for the Future.

Creswell, J. W. (2012). *Planning, conducting, and evaluating quantitative and qualitative research* (4th ed.). Pearson.

Darling-Hammond, L., Hyler, M. E., & Gardner, M. (2017). *Effective teacher professional development.* Learning Policy Institute.

Desimone, L. M. (2009). Improving impact studies of teachers' professional development: Toward better conceptualizations and measures. *Educational Researcher, 38*(3), 181–199. https://doi.org/10.3102/0013189X08331140

Ennis, R. H. (2016). Definition: A three-dimensional analysis with bearing on key concepts. In P. Bondy & L. Benacquista (Eds.), *Argumentation, objectivity, and bias: Proceedings of the 11th international conference of the Ontario society for the study of argumentation* (pp. 1–19). OSSA.

Fazio, X. (2009). Development of a community of science teachers: Participation in a collaborative action research project. *School Science & Mathematics, 109*(2), 95–107. https://doi.org/10.1111/j.1949-8594.2009.tb17942.x

Fazio, X., & Gallagher, T. L. (2018). Bridging professional teacher knowledge for science and literary integration via design-based research. *Teacher Development, 22*(2), 267–280. https://doi.org/10.1080/13664530.2017.1363084

Fazio, X., & Gallagher, T. L. (2019). Science and language integration in elementary classrooms: Instructional enactments and student learning outcomes. *Research in Science Education, 49*(4), 959–976. https://doi.org/10.1007/s11165-019-9850-z

Fernandes, G. W. R., Rodrigues, A. M., & Ferreira, C. A. (2020). Professional development and use of digital technologies by science teachers: A review of theoretical frameworks. *Research in Science Education, 50*(2), 673–708.

Goodnough, K. (2018). Addressing contradictions in teachers' practice through professional learning: An activity theory perspective. *International Journal of Science Education, 40*(17), 2181–2204. https://doi.org/10.1080/09500693.2018.1525507

Grierson, A. L., & Gallagher, T. L. (2009). Seeing is believing: Creating a catalyst for teacher change through a demonstration classroom professional development initiative. *Professional Development in Education, 35*(4), 567–584. https://doi.org/10.1080/19415250902930726

Guskey, T. R. (2002). Professional development and teacher change. *Teachers & Teaching, 8*(3), 381–391. https://doi.org/10.1080/135406002100000512

Guskey, T. R., & Yoon, K. S. (2009). What works in professional development? *Phi Delta Kappan, 90*(7), 495–500. https://doi.org/10.1177/003172170909000709

Hill, H. C., Beisiegel, M., & Jacob, R. (2013). Professional development research: Consensus, crossroads, and challenges. *Educational Researcher, 42*(9), 476–487. https://doi.org/10.3102/0013189x13512674

Hitchcock, D. (2018). Critical thinking. In E. N. Zalta (Ed.), *The Stanford encyclopedia of philosophy* (Fall 2020 ed.). https://plato.stanford.edu/archives/fall2018/entries/critical-thinking/

Hodges, C., Grant, M., & Polly, D. (2013, March). Beyond one-shot workshops: Three approaches to STEM teacher professional development. In *Society for information technology & teacher education international conference* (pp. 4795–4800). Association for the Advancement of Computing in Education.

Hotze, A., Gijsel, M., Vervoort, M., Peters, S., & Post, A. (2020). Preparing teacher educators for language-oriented science education: Design and impact of a professional development program. In *Edulearn20 proceedings* (pp. 636–643). IATED.

Krajcik, J. S., & Mun, K. (2014). Promises and challenges of using learning technologies to promote student learning of science. In N. Lederman & S. Abell (Eds.), *Handbook of research on science education* (Vol. 2, pp. 337–360). Routledge/Taylor & Francis.

Luft, J. A., & Hewson, P. W. (2014). Research on teacher professional development programs in science. In N. Lederman & S. Abell (Eds.,), *Handbook of research on science education* (Vol. 2, pp. 889–909). Routledge/Taylor & Francis.

Maeng, J. L., Whitworth, B. A., Bell, R. L., & Sterling, D. R. (2020). The effect of professional development on elementary science teachers' understanding, confidence, and classroom implementation of reform-based science instruction. *Science Education, 104*(2), 326–353.

McMillan, J. H. (2004). *Educational research: Fundamentals for the consumer* (4th ed.). Pearson.

Merriam, S. (2009). *Qualitative research: A guide to design and implementation.* Jossey-Bass.

Mutch-Jones, K., Hicks, J., & Sorge, B. (2022). Elementary science professional development to impact learning across the curriculum. *Teaching and Teacher Education, 112*, 103625. https://doi.org/10.1016/j.tate.2021.103625

National Academies of Sciences, Engineering, & Medicine. (2015). *Science teachers learning: Enhancing opportunities, creating supportive contexts.* National Academies Press. https://doi.org/10.17226/21836

National Research Council. (2013). *Next generation science standards: For states, by states.* National Academies Press. https://doi.org/10.17226/18290

National Science Teaching Association. (2020). *NSTA district professional learning packages.* https://www.nsta.org/nsta-district-professional-learning-packages

Ontario Ministry of Education. (2008). *Ontario curriculum grades 9 and 10: Science.* http://www.edu.gov.on.ca/eng/curriculum/secondary/science910_2008.pdf

Ontario Ministry of Education. (2022). *Science and technology.* https://www.dcp.edu.gov.on.ca/en/curriculum/science-technology

Organization for Economic Co-operation & Development. (2017). *The OECD handbook for innovative learning environments.* OECD Educational Research & Innovation. https://doi.org/10.1787/9789264277274-en

Owston, R., Wideman, H., Murphy, J., & Lupshenyuk, D. (2008). Blended teacher professional development: A synthesis of three program evaluations. *The Internet & Higher Education, 11*(3), 201–210. https://doi.org/10.1016/j.iheduc.2008.07.003

Penuel, W. R., Fishman, B., Yamaguchi, R., & Gallagher, L. P. (2007). What makes professional development effective? Strategies that foster curriculum implementation. *American Educational Research Journal, 44*(4), 921–958. https://doi.org/10.3102/0002831207308221

People for Education. (2019). *Connecting to success: Technology in Ontario schools.* https://peopleforeducation.ca/report/connecting-to-success-technology-in-ontario-schools/#chapter10

Peterson, A., Dumont, A., Lafuente, M., & Law, N. (2018). *Understanding innovative pedagogies: Key themes to analyse new approaches to teaching and learning.* OECD.

Polly, D., & Hannafin, M. J. (2010). Reexamining technology's role in learner-centered professional development. *Educational Technology Research & Development, 58*(5), 557–571. https://doi.org/10.1007/s11423-009-9146-5

Psillos, D. (2017). Development of a blended learning program and its pilot implementation for professional development of science teachers. In P. Anastasiades & N. Zaranis (Eds.), *Research on e-learning and ICT in education: Technical, pedagogical and instructional perspectives* (pp. 189–200). Springer. https://doi.org/10.1007/978-3-319-34127-9

Riggs, I. M., & Enochs, L. G. (1990). Toward the development of an elementary teacher's science teaching efficacy belief instrument. *Science Education, 74*(6), 625–637. https://doi.org/10.1002/sce.3730740605

Sandholtz, J. H., & Ringstaff, C. (2016). The influence of contextual factors on the sustainability of professional development outcomes. *Journal of Science Teacher Education, 27*(2), 205–226. https://doi.org/10.1007/s10972-016-9451-x

Sandoval, W. A. (2015). Epistemic goals. In R. Gunstone (Ed.), *Encyclopedia of science education* (pp. 1–6). Springer.

Sinclair, M., & Owston, R. (2006). Teacher professional development in mathematics and science: A blended learning approach. *Canadian Journal of University Continuing Education, 32*(2), 43–66. https://doi.org/10.21225/D52C75

Teddlie, C., & Tashakkori, A. (2009). *Foundations of mixed methods research: Integrating quantitative and qualitative approaches in the social and behavioral sciences.* Sage.

Thompson, J., Hagenah, S., Kang, H., Stroupe, D., Braaten, M., Colley, C., & Windschitl, M. (2016). Rigor and responsiveness in classroom activity. *Teachers College Record, 118*(5), 1–58. https://www.tcrecord.org, ID No. 19366.

Wayne, A. J., Yoon, K. S., Zhu, P., Cronen, S., & Garet, M. S. (2008). Experimenting with teacher professional development: Motives and methods. *Educational Researcher, 37*(8), 469–479. https://doi.org/10.3102/0013189x08327154

Wilson, S. M. (2013). Professional development for science teachers. *Science, 340*(6130), 310–313. https://doi.org/10.1126/science.1230725

Windschitl, M. (2009, February). Cultivating 21st century skills in science learners: How systems of teacher preparation and professional development will have to evolve. In *National Academies of Science workshop on 21st century skills*. https://sites.nationalacademies.org/cs/groups/dbasse site/documents/webpage/dbasse_072614.pdf

Zhu, C., & Wang, D. (2014). Key competencies and characteristics for innovative teaching among secondary school teachers: A mixed-methods research. *Asia Pacific Education Review, 15*(2), 299–311. https://doi.org/10.1007/s12564-014-9329-6

Xavier Fazio is a professor in the Department of Educational Studies, Faculty of Education at Brock University. Dr. Fazio's research broadly entails science and environmental sustainability education, teacher education and development, curricular innovation, instruction and assessment in science education. His research in Canada also connects to partnerships with researchers in the USA, European Union, Brazil, and Australia. Dr. Fazio's research has been supported with funding from the Social Sciences and Humanities Research Council of Canada (SSRHC), and various government agencies, industry, and educational associations. Currently, he is leading a multi-year Insight Grant from SSHRC examining how to connect school science to local communities using place-based perspectives to better promote meaningful engagement for science students. He currently is serving as Associate Editor for Journal of Science Teacher Education.

Kamini Jaipal-Jamani is a professor in Science Education in the Department of Educational Studies, Faculty of Education at Brock University. Her research and writing focus on science teaching and learning, technology integration, and teacher professional development. She has been involved in research partnerships with local school boards, professional teacher associations, and international organizations to support teacher and faculty professional development. Other professional activities include serving on the program committee of the Society for Information Technology and Teacher Education (SITE) and being a Lead Editor for the international journal, Cultural Studies in Science Education.

Chapter 8
Expanding Vision II of Scientific Literacy with an Indigenous Hawaiian Perspective

Poh Tan

8.1 Introduction

In this chapter I relate, through my personal teaching experiences and reflections, how I merged scientific and Indigenous perspectives to further extend my understanding of scientific literacy and learn to look at science more outwardly. I introduce a *Kānaka Maoli* (Indigenous Hawaiian People) perspective in science and later describe how this perspective guided and extended my science teaching experiences to consider other perspectives by taking "part confidently in discussions with others about issues involving science" (Twenty First Century Science, 2008, as cited in Roberts, 2010, p. 15).

Hereafter, the word *Indigenous* will be capitalized as a sign of respect to the community and to recognize Indigenous communities as part of the Canadian national identity. I acknowledge that there are many Indigenous worldviews, including 198 Indigenous nations in British Columbia (BC Government, 2022), where I currently live and work. Most Indigenous worldviews see "humans having a seamless relationship with nature which include seas, land, rivers, mountains, flora, and fauna" (Cunningham & Stanley, 2003, p. 403). The unification of the human community with the natural world is part of the traditional knowledge of "'indigenous' peoples the world over, whether Māori, Hawaiian, African, Native American and so on" (Royal, 2002, p. 29).

The subject of science was introduced in European school curricula in the early nineteenth century partly due to supporters and advocates of science teaching such as Thomas Huxley, Herbert Spencer, John Tyndall, Michael Faraday, Charles Lyell, and Charles Eliot. At the time, "the humanities were firmly entrenched as the subjects that

P. Tan (✉)
Faculty of Education, Simon Fraser University, University Drive, Burnaby,
BC 8888V5A 1S6, Canada
e-mail: pctan@sfu.ca; poh@stemedgeacademy.com

were thought to lead to the most noble and worthy educational outcomes" (DeBoer, 2000, p. 583). Science was defended as a legitimate intellectual study based on "the power it gave individuals to act independently" (p. 583), and science education was justified for "its relevance to contemporary life and its contribution to a shared understanding of the world on the part of all members of society" (p. 583). By the 1930s, there were concerns that science education focused primarily on content and needed to instead focus on a broader perspective of science that would encompass "personal development … to help individuals adjust to life in modern society" (p. 586). The shift from learning science content knowledge to learning how to critically apply that knowledge became prominent after the launch of Sputnik into Earth's orbit by the Soviet Union in 1957 (Roberts & Bybee, 2014). At that time, the terms *science literacy* and *scientific literacy* began to emerge in the academic literature.

Science literacy was first introduced by Hurd in his 1958 paper *Science Literacy: Its Meaning for American Schools*. Science literacy is often used interchangeably with scientific literacy. Roberts (2007) described two visions of scientific literacy, labelling them Vision I and Vision II. Vision I suggests that a scientifically literate individual understands science from the perspectives of content and processes, a way of looking "inward at science, to build curriculum from its rich and well-established array of techniques and methods, habits of mind, and well-tested explanations for the events and objects of the natural world" (Roberts & Bybee, 2014, p. 546). Vision II suggests that a scientifically literate person understands science from a broader perspective and can apply science in situations with a scientific component, considering other perspectives, and including the scientific perspective to make informed decisions about society (Roberts, 2010). In other words, Vision II proposes that science curricula include "how science permeates and interacts with many areas of human endeavor and life situations" (Roberts & Bybee, 2014, p. 546). Science curriculum development can be characterized by a consistent competition of both visions of the end goal of school science and that the root of this competition stems from the interplay between looking at science inwardly or outwardly (Roberts, 2011).

To date, there is no gold-standard definition of a scientifically literate person or scientific literacy. Eshach (2006) outlined science education approaches with primary and preschool children that encourage and "nurture scientific thinking skills and inculcate in children the desire and passion to know and learn" (p. xii). Scientific inquiry is a continual and messy process that involves questioning, observing, and testing in a non-systematic way within a scientific context; it is through these messy ways of inquiring that student scientific learning and teacher reflections begin to take shape (Bybee, 2002). However, it is not akin to a lockstep scientific method but does involve processes such as identifying problems, forming hypotheses, conducting experiments, analyzing data, and sharing findings. For example, in BC, the K–12 science curriculum is inquiry based and intended to support students in critical thinking and problem solving, ethical decision making, and scientific communication of question and ideas (BC Ministry of Education, 2021). Although the current curriculum includes guidance toward teaching a Vision II conception of science, lessons are often designed from a Vision I perspective when constrained by materials and time.

8.1.1 Personal Reflection—Visions I and II

As I reflected on my previous experience as a research scientist in the field of stem cells, I realized that I was trained to think about science primarily within Vision I. My colleagues and I followed the unspoken, but commonly understood, scientific tradition that the practice and application of science originates from a place of knowledge, with related technical processes, grounded in pure objectivity. Although we reflected on our mistakes from failed experiments in the laboratory, we often did not appreciate the consequences and effects of that work beyond the preclinical laboratory or medical field. Specifically, in my work, I considered how genetically modified retroviruses could be used to understand mechanisms in cellular mobilization for therapy in addition to indirect infections and gene delivery of harmful sequences to human systems. Although I often thought about and practiced science from a Vision 1 perspective, I did not intentionally engage in a Vision II perspective.

When I became a novice science educator, I frequently designed lessons that embodied characteristics of Vision I; oftentimes, I felt it was most important to not deviate from a lesson plan designed to ensure the students met specific learning outcomes. In other words, I saw myself as an educator who knew the right pedagogy for my class and that deviation from a set lesson plan meant my students would not meet my teaching objectives. I realized my bias and started to reflect on my teaching practices while teaching a group of preschoolers about the anatomy of dinosaur teeth; I share this reflection in the next section.

In my pursuit of a second doctoral degree, this time in the field of education, I became aware of the need for thinking about how to apply science knowledge in different ways, for example, taking a philosophical and/or contemplative stance as opposed to a purely laboratory-based stance. I now recognize that Vision I and Vision II are not mutually exclusive but instead are interdependent.

8.1.2 Personal Reflection—An Expanded Vision II (with Indigenous Perspectives)

As I reflected on my science teaching in a preschool classroom, I realized that I often started with preplanned activities. I remember walking into a classroom with 10 children under the age of 5 years old. My goal was to teach the differences between herbivores and carnivores by looking at dinosaurs. The children gathered information about what type of food was eaten by observing the shape of the dinosaur's teeth. One child raised his hand—while his other hand was exploring the teeth in his mouth—and asked, "Ms. Poh, our teeth don't look like that, how come? I eat chicken, I don't like carrots! Carrots yucky." The other children repeated with giggles, "Ya! Carrots yucky! Carrots yuck!" After that moment, I quickly put away my plastic dinosaur models and prompted the children to use mirrors to look at their own teeth.

I realized that I had been placing more emphasis on transmitting scientific knowledge—insisting that the main learning objective was children knowing the structural differences between herbivores and carnivores—than being open to the possibilities for children's inquiry. In this reflection, the question from one child about their own teeth caused me to pause, listen to the child, and initiate an organic inquiry because I recognized the importance of the child's perspectives and contributions. From this point forward in my teaching, I was more aware of the questions that children were asking and how they interacted with materials, the environment, and me during a science lesson.

At the 2016 Canadian Society for the Study on Education conference, I attended a presentation where it was proposed that science literacy could be considered from a perspective that integrates multiple perspectives about scientific worldviews: a system that embeds Indigenous systems of knowing and non-Western paradigms (Murray, 2016). Lederman (2006) noted that "science is a human enterprise ... [and, therefore, science] is affected by the various elements and intellectual spheres of the culture in which it is embedded" (p. 306).

Western paradigms situate the scientist as separate from the object of their study. In contrast, the Indigenous worldview places "special significance or weight behind the idea of the unification of the human community with the natural world" (Royal, 2002, p. 29). Indigenous communities traditionally have strong connections to the land, interacting with local ecosystems with a deep spiritual purpose as they rely on the land to meet their needs (Cunningham & Stanley, 2003; Pascua et al., 2017). In Indigenous worldviews, generally "there is no conceptual separation between the spiritual and natural world, which makes their cultural worldview conceptually and symbolically different from Western thinking" (Tyler, 1993, p. 227). Separating knowledge about the natural world from its context, as a Western perspective of science education does, is inadequate for Indigenous and Western populations alike.

The point is that science education can be strengthened by incorporating an Indigenous worldview. For example, Kealiikanakaoleohaililani et al. (2018) argued that taking a spiritual approach to the natural environment advanced "the science of sustainability, the management of natural resources, and the conservation of nature" (p. 2) with individuals who were more effective stewards because of their enhanced ability to interact with their environment. Similarly, Snively and Williams (2016) pointed out that when a child speaks to a tree the child is "engaged in coming to know the connections of the universe and to feel empathy with another living entity" (p. 39).

8.1.3 The Kānaka Maoli Worldview

Drawing on this idea about the benefits of complementary worldviews, I pictured an expanded Vision II of science literacy that braids Indigenous and Western worldviews and is built upon relationships between and within humans and nonhumans and between nature and culture. Here, I focus on the Kānaka Maoli worldview that helped me to build a deeper relational connection between my research and practice as a scientist and my science teaching practices. I describe how the stories and *hula* of Kānaka Maoli challenged me to reconceptualize how I teach science by introducing an approach that moves toward stewardship of the Earth.

Hawaiian epistemology is an "ocean of knowing" (Meyer, 2001, p. 126) and an "embodiment of oceanic knowledge" (Martinez, 2021, p. 392) that can be experienced through spirituality and knowing, physical place and knowing, cultural nature of the senses, relationship and knowledge, utility and knowledge, words and knowledge, and the body/mind question. Here I discuss relationship and knowledge in depth as they apply to teaching young children scientific concepts. I acknowledge that the other experiences are important contributions toward relationality and connection to the world and form a holistic approach to knowing and understanding that creates a deeper connection with oneself and land. For the scope of this chapter, I reflect on how forming deep relationships with land and others can help shift one's views and actions in the world through science.

In 2004, I was introduced to the history, culture, *moʻolelo* (stories), *mele* (songs), and *oli* (chants) of Hawaiʻi when I attended a hula class at a *kumu* (master) hula *halau* (school). Hula is Hawaiʻi's dance that connects history, moʻolelo, *Aloha*, heart, and spirit to land, people, and ancestors. Hula is a process of self-discovery to allow an emergence and growth where knowledge, understanding, and knowing is connected through our body, heart, mind, and spirit to form deep connections with *ʻaina* (world, land, ocean, sky). "Hula is a moving encyclopedia inscribed into the sinews and postures of dancers' bodies. It carries forward the social and natural history, the religious beliefs, the philosophy, the literature, and the *scientific knowledge* [emphasis added] of the Hawaiian people" (Rowe, 2008, p. 31). Furthermore, "hula becomes an entryway into a series of integrated, unfolding experiences and layers of meaning through which knowledge deepens and broadens as it is explored" (Rowe, 2020, p. 139). The heart of my halau and the intention of my kumu are to share beautiful but meaningful mele and oli. My kumu, Josie de Baat, often reminded us, "the most important thing of dancing the hula is to know, learn, and embrace Hawaiian culture, history, and moʻolelo with your whole heart, and most important, with Aloha" (de Baat, personal communication, March 29, 2019). The word Aloha means more than a greeting and instead is a practice and way of being that shapes our relations with each other and the world (Ebersole & Kanahele-Mossman, 2020). The culture, history, and moʻolelo includes science:

> Hawaiians were excellent scientists before Western contact, as shown by their abilities and practices that are only now being (re)discovered by others around the world ... [for example,] their use of observation, theory development, and hypothesis testing in agriculture. (Allaire, 2013, p. 37)

In the Kānaka Maoli worldview and as in Indigenous worldviews generally, knowledge is relational and incorporates human interactions with and in nature that involve "body, mind, soul, and spirit with all aspects of nature" (Cajete, 2000, p. 64). Interconnectedness, relationality, and reciprocity move beyond a body of knowledge for Indigenous people and instead is a way of life (Snively & Williams, 2016). Hawaiians "practice reciprocity, exhibit balance, develop harmony with land, and generosity with others" (Meyer, 2001, p. 134) as they draw on relationships to shape and share knowledge. My practice of hula informs and shapes my identity as a scientist and science educator. Hula is one way that the Kānaka Maoli relational connection with the world is preserved for the next generation through education, including science education.

Relationship is often missing in the teaching of science; the object of study is frequently removed from its place thereby creating a relational distance (Kealiikanakaoleohaililani et al., 2018). When educators are teaching young children about science, it is important to emphasize (a) children's connections to science knowledge and understanding and (b) how those connections are reciprocal. In taking an expanded Vision II perspective, I am shifting away from teaching science as content and skills toward teaching science to include relationality, particularly the connection between self and land. Being intentional about relationality in science goes beyond merely taking our students for a walk in the forest; it requires purposeful and deliberate engagement with a focus on interconnections.

Teaching science with a Kānaka Maoli lens may help to eliminate a dualistic and objectified worldview, thus leading to a deeper understanding of our place in the natural world. The connection that can be acquired through hula can help deepen a realization of our interdependence; with *kuleana* (responsibility), a reciprocal relationship is understood. The following case study is an example of how I used a relational Indigenous Hawaiian perspective to create an expanded Vision II of scientific literacy and to guide my teaching practice with a small group of preschool children.

8.2 Creating Relationships Through a Biome Project with Preschool Children

The case study was a 5-month teaching-research project at a preschool for children aged 2½ to 6 years old in Richmond, British Columbia. Trinity's Little Children Centre was located at a church within an urbanized area with a high Asian population, surrounded by concrete sidewalks and asphalt roads, with a lack of access to green space. The outdoor space at the preschool was part of a parking lot enclosed with a chain-link fence; it was comprised of a small gravel area, artificial grass, and three large planters with small evergreen shrubs. In this area, the children had access to toys, a small plastic slide, and some tricycles. The indoor space included a small gym and a classroom that accommodated 12–20 preschoolers.

The objective of the preschool's curriculum was to guide children in learning about their interconnectedness with God and how this relationship contributes to respect and reciprocity with Earth and all living things. The preschool followed a play-based curriculum where children were provided independent and teacher-guided explorations with a combination of free-play and planned learning activities. Daily programs were determined one month in advance and included activities such as centre-time with building blocks, puzzles, and arts and crafts. Weekly Mandarin lessons were provided. Using the BC curriculum (BC Ministry of Education, 2021) as a guide, science lessons were typically taught twice a week. For two months before the research project started, I visited the preschool to learn, listen, and understand the community. Once or twice a week for 60–90 min, I helped with snack preparation, reading time, and outside play to familiarize myself with daily routines and to establish relational and reciprocal trust with the children. I was another adult with whom the children could chat. I answered their questions about my presence and my intentions; more importantly, I gained the knowledge and understanding that would enable me to respect each child's learning identity.

In preparation for teaching, I had multiple face-to-face meetings and email conversations with Ms. Eva To, the lead educator, to ensure that the lessons and activities aligned with the curriculum, her programming intentions, and the preschool's objectives. The lead educator, who was a participant in this case study, was a certified early childhood educator with over 30 years of experience in Hong Kong and in Canada. Her teaching philosophy was Reggio-Emilia inspired, meaning she viewed children as social learners who are co-constructors of knowledge (Hewett, 2001). Furthermore, the Reggio-Emilia Approach positions the educator as "collaborator and co-learner along with the child, a guide and facilitator" (p. 95). Twelve children aged 4–5 years were participants in the study (see Fig. 8.1, permissions granted for all photographs).

To collect data, the lead educator and I took pictures with our phones, and I documented conversations through a combination of notetaking and audio recording. At the beginning of each visit, I asked the children if they assented to being photographed

Fig. 8.1 Lead educator Ms. Eva To (left) and the author with the participating children (right)

or recorded. On some days, some children preferred not to be recorded; in those sessions, I would take only notes and pictures.

For my case study, the lead educator and I decided on the science topic then collaboratively designed and taught the lesson. Typically, some science topics would be selected by the lead educator based on the children's interests and the time of year (e.g., winter was a topic in November and December so discussion focused on the cycle of seasons, hibernation, conifers, and snow); other science topics were based upon daily routines in the classroom (e.g., hand washing, cleaning up, diet, and exercise). Both of us agreed that an appropriate topic for spring was biomes.

We designed the biome unit to show the children different environments and how they were different from their everyday surroundings. We discussed the lack of green space around the preschool and how the children did not frequently visit natural places such as parks and the seashore. I wanted the children to learn about how different types of plants adapt to different environments and how specific biomes contribute to the health of our air and water and to know that distant environments are important to us. For example, we talked about how volcanoes in Hawai'i helped shape the surface of the Earth, created the sea floor, and impacted the air we breathe even though they are far away. The culminating activity would be an arts-based project where the children build their own biome and express their individual and collective personal perspectives of their relationships with those biomes (e.g., what they knew and what they felt) in a combined representation.

8.2.1 Structure of the Lessons

Each weekly lesson lasted between 45 and 90 min. We began with a discussion of what the children had learned the previous week and an introduction to the day's topic. After the discussion, I gave them a choice of a story or a song. Next, we would typically use our senses to explore aspects of biomes. Finally, they would be given the opportunity to work on our culminating collaborative arts-based endeavour—a biome diorama.

The discussion typically lasted about 10 min, and the focus was mostly situated within Vision I—content knowledge. For example, in one lesson we talked about how a turtle is different from a tortoise and how each part of the animal's body adapted to its environment. We often moved our bodies to reinforce the discussion topics (e.g., moving like turtles and tortoises, mimicking seaweed). In some lessons, the children had many questions and the discussion continued for up to five extra minutes. The discussions were conversational in nature, and I followed their interests as much as possible.

After the discussion, I would continue our exploration of the lesson's topic through story or song. If the children chose a story, the lead educator or I might read a book that we had preselected. During reading, we would focus on each page, asking questions about the different biomes depicted and about the plants and animals that lived within them. Sometimes I created a story to connect and relate to different biomes that they were learning. I made notes for these stories but did not write them out in full because I wanted the storytelling to be as fluid as possible. For example, the following bullet points were my notes for a volcano story:

- When volcanoes erupt, they create land; link back to the Lava song about how the eruption by Uku made him rise up in the sea.
- The environment around the eruption is really hot and has high concentrations of sulfur but animals still live there.
- After Lele erupted and rose above the sea, the smoke from her eruption made ashes.
- Ashes from volcano can help plants grow.
- Different types of plants grow on the sides of volcanoes.

In the storytelling sessions, I was driven by questions the children had; I pivoted quite a bit but at the same time stayed within the topic of the lesson.

If the children chose a song, the lead educator and I would introduce a new song about plants, animals, and/or biomes or we would repeat a song that they already knew. Sometimes we would make up a song; we would follow the children's lead with the tunes they hummed and the words they chose. In one lesson, I brought my ukulele. I had been learning to play the song Lava, which is a geological love story about underwater volcanoes, eruptions, and extreme environments. The song describes how two volcanoes, Uku and Lele, rise, appear, disappear, and reappear. I played my ukulele and sang with the children to evoke images and discussions about how volcanoes are seen as family and spiritual connections in Hawai'i. While singing songs, we sometimes moved our bodies to mimic the words and our understandings of biomes, for example, to move like ocean currents, seaweed, turtle flippers, and volcanic eruptions.

After a song or story, we would explore different specimens (e.g., leaves, seeds, pinecones, roots, flowers) or listen to different animal sounds. For example, in one lesson we discussed how plant physiology varies depending upon the biome, specifically terrestrial versus aquatic plants. We had already collected specimens from local shores (e.g., aquatic plants such as seaweed) and gardens (e.g., terrestrial plants such as ferns). We observed the specimens, using our sense of touch to explore them more closely (Fig. 8.2). In some cases, it was the first time some children had touched the various specimens. As we examined them, we talked about the importance of each specimen and how it contributed to each biome. The terrestrial plant specimens guided the children toward asking questions about the health of the plant, soil, and environment. Their questions about the absence of biomes and certain plants conveyed their concern for human and animal health. Their dialogue about how they had experienced these specimens and how they noticed differences and similarities between more natural biomes and urbanized areas in the cities continued throughout the lesson and even between my visits to the classroom.

Fig. 8.2 Children exploring specimens from terrestrial biomes

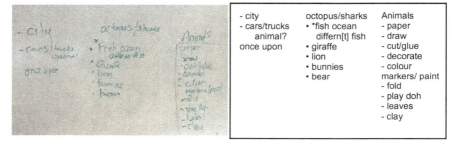

- city	octopus/sharks	Animals
- cars/trucks	• *fish ocean	- paper
animal?	differn[t] fish	- draw
once upon	• giraffe	- cut/glue
	• lion	- decorate
	• bunnies	- colour
	• bear	markers/ paint
		- fold
		- play doh
		- leaves
		- clay

Fig. 8.3 Results of a brainstorming session for the culminating arts-based biome project

8.2.2 The Arts-Based Project—A Biome Diorama

At the end of each lesson, children could contribute to the biome diorama. There were four designated tables set up in the classroom with different materials available where they could work independently or collaboratively. Children could choose to work at a specific table, go back and forth between tables, or work at the big table with the large cardboard base that would eventually be the completed diorama. They would choose a biome either from our lessons or the environments that they experienced every day (e.g., playgrounds, walkways, beaches). I led brainstorming sessions about what would and would not belong in a biome; they identified different animals, characteristics, or features that they would like to include. They would ask questions such as, "Where are the animals in this biome? Where are the bears? Where are the fish? Where are the birds?" Fig. 8.3 shows the chart made during one brainstorming session.

During this brainstorming session, one child was curious about the absence of cars, trucks, and cities in our planning for the diorama, as shown in the following excerpt.

David: Ms. Poh, why does the biome have no cars, no trucks, and no city?
Me: I like your question, David; why do you think a biome needs those things?
David: I want to add a truck and car and a road because we can drive there and …
         [David was interrupted by Eva]
Eva: Why do you want to add a car and road? That is not good for the animals.
         The animals can die and the trees can die and … [Eva was interrupted by David]
David: No, we need a car and a truck to drive there. We don't kill the animals.
         [David looks curiously at Eva, mumbles a statement in Mandarin]
Me: David, let's listen to what Eva thinks. Eva, what do you think of David's question about the animals?
Eva: No, Ms. Poh. David! The cars are not good for the animals because the biomes don't have cars, only animals, trees, fish, and ocean. The car and truck are poison.
Me: That's a very good observation, Eva; in all the examples of biomes we have learned about, there are no cars, trucks, or roads. Biomes are natural environments where animals and plants live. You are right that the car and truck may not be good for the animals.

We also talked about the materials we needed to build the biomes and their features; we wanted to avoid using plastic. The children, the lead educator, and I had previously collected recycled and natural materials from our homes and during trips to parks and beaches. We learned about the role and importance of each material; for example, female pinecones are reproductive structures from pine trees that, in addition to helping trees propagate, are part of animals' food chains. The children incorporated the collected materials as they constructed their biomes (Fig. 8.4); for example, one child made a crab from rocks, small pinecones, and red maple seeds.

The biome diorama took shape over the 17 weeks of the research project as shown in Table 8.1. Sometimes, the lead educator allowed the children to work on the diorama on days when I was not at the preschool. On my next classroom visit, I could tell that the diorama had grown; there would be more sand on the beach or more colour on the mountains.

8.3   Discussion

During this research project, I was influenced by the Kānaka Maoli worldview and my involvement with kumu hula Josie as I adopted an expanded Vision II of scientific literacy and worked to establish a respectful learning environment in which I was guided by the children's interests and their questions. My awareness of the importance of the children's contributions led me to ensure that they had multiple opportunities

Fig. 8.4 Creating individual biomes and their features for the diorama

to make choices as they explored the topic of biomes and created the diorama. For example, although I always had lesson plans prepared, I also gave them choices in selecting specific activities or discussion topics and in how they would participate in building the diorama.

In some contexts, such as introductory discussions and specimen examinations, our biome unit emphasized a Vision I perspective of scientific literacy. For example, when we identified the science content that we wanted the children to learn, we were following a Vision I approach to science education. In other contexts, such as our discussions during nature walks, we emphasized a Vision II perspective. For example, when we focused on extending the children's understanding of biomes to think about how biomes are important to life situations, we were following a Vision II approach. When we incorporated a focus on their relational connections (e.g., to God, in various biomes, with one another), we were expanding Vision II to include an Kānaka Maoli approach. For example, we discussed aspects of relationships and stewardship as we explored the role of biomes in supporting humans and nonhumans. However, most

Table 8.1 Progression of the biome diorama

Timeframe and focus	Photo of diorama at end of timeframe
Week 1: General Discussion To begin our biomes exploration, I asked the children what they knew about different types of places. They were asked to draw what they thought were different types of biomes or habitats. We talked about creating a diorama and what that process might look like. They were asked to think of natural and recycled materials that we could use to create our diorama	
Weeks 2–3: The Forest The children were provided with natural and recycled materials to create trees and the forest floor. The forest was made from coloured paper, and trees were cut from recycled cardboard. The forest floor was made from moss, rocks and stones, dried leaves, and pinecones. They decided that the forest needed to be in the centre of the diorama amongst all the biomes that we would add because it would help connect the mountains, rivers, and oceans	
Weeks 4–6: The Ocean and Rivers While constructing the ocean and rivers, the children talked about how some rivers come from mountains and end in the ocean. A river connects different biomes and creates an environment that supports plant and animal life. They reflected on how water is different between a river and the ocean. They painted the ocean area and then added recycled coloured paper to create a 3-dimensional wave effect. They used natural materials such as rocks, sand, and leaves to create the riverbank and paint to create the water	
Weeks 7–10: The Mountains and the Seashore It was a 2-week project to create the mountains using recycled materials such as paper cups and paper. The texture from the paper represents the topography of the mountain. Making mountains with glue got messy but was so much fun! The children collaborated on building a beach with natural materials such as sand, seashells, and rocks. As they created the beach, they talked about how they loved playing in the sand and the waves	
Weeks 11–14: Details The details the children added to the diorama included many kinds of animals that were made mainly with natural materials. They used rocks, pinecones, twigs, and maple seeds to create a crab for the seashore. They made bears for the mountains with different types of beans. They made different types of fish for the ocean and rivers using beans, sequins, and seeds	

(continued)

Table 8.1 (continued)

Timeframe and focus	Photo of diorama at end of timeframe
Weeks 15–17: Clouds, Birds, and More Details During construction, the children decided to add birds because they were concerned that "the birds will feel bad because they were not included in the project." Two children suggested that we needed to add an atmosphere to the mountain range with clouds for the birds to fly through. As we made clouds from paper and cotton balls, we learned about how important it was to keep our atmosphere clean. They added more geological features to the diorama (e.g., rocks, sand) and a few more animals	

of the time our learning together was co-constructed within the intertwined spaces of Visions I and II and an expanded Vision II.

In my teaching during the biome project, I worked alongside the children who collectively and collaboratively learned through song, dance, and story. My teaching practices were informed by my experiences as a student of hula, which brought an awareness of the importance of building deep relational connections while constructing scientific understanding. I planned each lesson to intentionally reinforce the connection between scientific knowledge and cultural perspectives. By incorporating an Indigenous epistemology, I could emphasize relationality and address the critical role of stewardship in our interconnected world.

Hawaiian epistemology positions relationships as foundational in the construction of knowledge (Meyer, 2001). During this project, each conversation, interaction, song, dance, or story, whether planned or spontaneous, helped the children build relationships with other people and with the biomes. The diorama provided an extended opportunity to express themselves and their understanding of biomes through finger painting and working with dirt, sand, moss, and glue; these hands-on experiences strengthened their relationships with the materials and with the biomes they were creating. Building the diorama also allowed the children to experience a sense of working together—with me, with the lead educator, and with each other—similar to how they understood each biome's interdependence. Learning about our relationships to others, including volcanoes, animals, and plants, helped the children create a spiritual connection.

As they worked on the diorama, the children talked about their relationships with and, more importantly, their responsibilities for each biome. For example, one child built a "garbage basket so that everybody can put their garbage there when they go on the sand [the beach]." They talked about their experiences seeing mountain ranges, visiting beaches, hiking on trails, and driving inside parks (i.e., paved roads located inside large parks); but ultimately, they decided not to add cities, cars, trucks, or roads to the diorama. Collectively, they felt that those features did not belong because of their potential negative affects on the environment. The sense of stewardship felt by these children was clear.

8.4 Concluding Remarks

In this chapter, I have shared my journey from scientist with a strong Vision I perspective of scientific literacy to science educator following an expanded Vision II perspective and described how I was influenced by the Kānaka Maoli worldview. Vision I of scientific literacy focuses on the content and processes of science itself. Vision II takes a broader perspective and focuses on the application of science in making evidence-based decisions about societal issues. An expanded Vision II incorporates Indigenous worldviews and practices and decenters the Western paradigm that typically dominates science education. The Kānaka Maoli approach to relating to the world has helped me to practice a different relationship to science through hula, which challenged my thinking about science teaching.

My practice as a hula haumana informs and shapes my identity as a scientist and science educator who aspires to build deeper connections and teach from a place of relationality. For the Kānaka Maoli, aina shapes and influences understanding and is "an epistemological cornerstone to our ways of rethinking [that] is all about relating in ways that are sustaining, nourishing, receptive, wise" (Meyer, 2008, p. 5). In my current practice as a science educator, I view scientific literacy as the scientific knowledge and understanding required to make responsible choices for personal well-being and for the well-being of others, including nonhuman entities. Drawing on a Hawaiian worldview where knowledge is relational and reciprocal, I experienced a deeper awareness of the connections that can be fostered through science education from an expanded Vision II perspective.

Acknowledgements The author would like to acknowledge and thank the following individuals and organizations for supporting this research: Kumu Hula Josie de Baat, Dr. Margaret MacDonald, Ms. Eva To, Ms. Jade Leong, Simon Fraser University, Faculty of Education, the children at the preschool, Faculty of Education Graduate Fellowship, and peers and colleagues for their continual mentorship and support.

References

Allaire, F. S. (2013). Navigating rough waters: Hawaiian science teachers discuss identity. *Educational Perspectives, 46*(1 & 2), 31–39. Retrieved from https://coe.hawaii.edu/publications/educational-perspectives-2014/

British Columbia Government. (2022). *B.C. first nations & indigenous people.* https://www.welcomebc.ca/Choose-B-C/Explore-British-Columbia/B-C-First-Nations-Indigenous-People#:~:text=They%20include%20First%20Nations%2C%20Inuit,own%20unique%20traditions%20and%20history

British Columbia Ministry of Education. (2021). *BC's curriculum.* https://www.curriculum.gov.bc.ca/curriculum

Bybee, R. W. (ed.). (2002). *Learning science and the science of learning: Science educators' essay collection.* NSTA Press.

Cajete, G. A. (2000). *Native science: Natural laws of interdependence.* Clear Light.

Cunningham, C., & Stanley, F. (2003). Indigenous by definition, experience, or world view [Editorial]. *British Medical Journal, 327*(7412), 403–404. https://doi.org/10.1136/bmj.327.741 2.403

DeBoer, G. E. (2000). Scientific literacy: Another look at its historical and contemporary meanings and its relationship to science education reform. *Journal of Research in Science Teaching, 37*(6), 582–601. https://doi.org/10.1002/1098-2736(200008)37:6%3c582::AID-TEA5%3e3.0.CO;2-L

Ebersole, M., & Kanahele-Mossman, H. (2020). Broadening understandings of the cultural value of Aloha in a teacher educator program. *Journal of Culture and Values in Education, 3*(2), 81–99. https://doi.org/10.46303/jcve.2020.14

Eshach, H. (2006). *Science literacy in primary schools and pre-schools.* Springer.

Hewett, V. M. (2001). Examining the Reggio Emilia approach to early childhood education. *Early Childhood Education Journal, 29*(2), 95–100. https://doi.org/10.1023/A:1012520828095

Hurd, P. D. (1958). Science literacy: Its meaning for American schools. *Educational Leadership, 52*, 13–16 & 52.

Kealiikanakaoleohaililani, K., Kurashima, N., Francisco, K., Giardina, C., Louis, R., McMillen, H., Asing, C., Asing, K., Block, T., Browning, M., Camara, K., Camara, L., Dudley, M., Frazier, M., Gomes, N., Gordon, A., Gordon, M., Heu, L., Irvine, A., …, & Yogi, D. (2018). Ritual + sustainability science? A portal into the science of Aloha. *Sustainability, 10*(10), 3478–3495. https://doi.org/10.3390/su10103478

Lederman, N. G. (2006). Syntax of nature of science within inquiry and science instruction. In L. B. Flick & N. G. Lederman (Eds.), *Scientific inquiry and nature of science: Implications for teaching, learning, and teacher education* (pp. 301–317). https://doi.org/10.1007/978-1-4020-5814-1_14

Martinez, E. B. (2021). Review of the book 'Waves of knowing: A seascape epistemology', by K. Ingersoll. *Island Studies Journal, 16*(1), 392–393.

Meyer, A. M. (2001). Our own liberation: Reflections on Hawaiian epistemology. *The Contemporary Pacific, 13*(1), 124–148. https://doi.org/10.1353/cp.2001.0024

Meyer, A. M. (2008). Indigenous and authentic: Hawaiian epistemology and triangulation of meaning. In N. K. Denzin (Ed.), *Handbook of critical and Indigenous methodologies* (pp. 217–232). SAGE.

Murray, J. J. (2016, May 28–June 1). *Science education in Canada: Toward a new orientation to the sustainability sciences* [Paper presentation]. Canadian Society for the Study of Education Annual Conference, Calgary, AB, Canada. https://ocs.sfu.ca/csse/index.php/csse/CSSE2016/paper/viewFile/2928/38

Pascua, P., McMillen, H., Ticktin, T., Vaughan, M., & Winter, K. B. (2017). Beyond services: A process and framework to incorporate cultural, genealogical, place-based, and indigenous relationships in ecosystem service assessments. *Ecosystem Services, 26*(Part B), 465–466. https://doi.org/10.1016/j.ecoser.2017.03.012

Roberts, D. A. (2007). Scientific literacy/science literacy. In S. K. Abell & N. G. Lederman (Eds.), *Handbook of research on science education* (Vol. I, pp. 729–780). Lawrence Erlbaum Associates.

Roberts, D. A. (2010). Competing visions of scientific literacy: The influence of a science curriculum policy image. In C. Linder, L. Östman, D. A. Roberts, P.-O. Wickman, G. Ericksen, & A. MacKinnon (Eds.), *Exploring the landscape of scientific literacy* (pp. 11–27). Routledge. https://doi.org/10.4324/9780203843284

Roberts, D. A. (2011). Competing visions of scientific literacy: The influence of a science curriculum policy image. In E. Gaalen, R. A. Douglas, P. Wickman, L. Östman, C. Linder, & A. MacKinnon (Eds.), *Exploring the landscape of scientific literacy* (pp. 11–27). https://doi.org/10.4324/978020 3843284

Roberts, D. A., & Bybee, R. W. (2014). Scientific literacy, science literacy, and science education. In N. G. Lederman & S. K. Abell (Eds.), *Handbook of research on science education* (Vol. II, pp. 545–558). Routledge.

Rowe, S. (2008). We dance for knowledge. *Dance Research Journal., 40*(1), 31–44. https://doi.org/10.1017/S0149767700001352

Rowe, S. M. (2020). Where our feet fall: A hula journey into knowledge. *Pacific Asia Inquiry*, *11*(1), 134–150. https://www.uog.edu/_resources/files/schools-and-colleges/college-of-liberal-arts-and-social-sciences/pai/v11/15_pai11_rowe.pdf

Royal, C. T. A. (2002). *Indigenous worldviews: A comparative study* [Report]. Te Wānanaga-o-Raukawa. https://static1.squarespace.com/static/5369700de4b045a4e0c24bbc/t/53fe8f49e4b06d5988936162/1409191765620/Indigenous+Worldviews

Snively, G., & Williams, W. L. (Eds.). (2016). *Knowing home: Braiding Indigenous science with western science* (Book 1). University of Victoria (CC-BY-NC-SA 4.0).

Tyler, M. E. (1993). Spiritual stewardship in aboriginal resource management systems. *Environments, 22*(1), 1–8.

Poh Tan BSc (Simon Fraser University), PhD (University of British Columbia), PhD ABD (Simon Fraser University), is a stem cell biologist, entrepreneur, mother, and graduate student currently completing her thesis research on understanding scientific literacy with an Indigenous Hawaiian framework and hula. After obtaining her first PhD in stem cell biology, she worked in the biotechnology industry that helped developed her entrepreneurial spirit. She is founder of STEMedge Academy Inc. and a mother to two young boys who inspired her to pursue a doctorate in science education. Her research interest is in science education that seeks to decenter a dominant way of teaching science, specifically through the practice of hula. She is of Nyonya descent from Malaysia and she is inspired everyday by her children, the students she shares knowledge with, and the sprit of Aloha.

Chapter 9
Using Bee-Bots® in Early Learning STEM: An Analysis of Resources

G. Michael Bowen, Eva Knoll, and Amy M. Willison

9.1 Introduction

On a late afternoon I walked into a daycare, where I was friends with some staff, carrying a Bee-Bot® by my side … and I sat down at a kid-high table. Several of the 3- and 4-year-olds I'd talked with many times over the previous months ran over and sat down to say hi to "Bear" (my nickname) and started telling me about their day, talking excitedly over each other as they smiled and laughed. I picked up the Bee-Bot and set it on the table. Most of them stopped talking for a few seconds, and then they erupted with questions. While they were talking, and as they were watching me and the Bee-Bot intently, I pressed the button that erased previous programs and pressed the forward button twice, the right turn button once, and then the forward button again. It beeped each time; and after the first beep, many of them stopped talking and watched. I then pressed the green "Go" button; the Bee-Bot moved forward and beeped and the eyes blinked, and the children exploded with joyful sound, comments, and questions all at once. More children clustered around as the Bee-Bot kept moving. The Bee-Bot stopped for a final time, beeped and flashed multiple times; I then reached over and picked it up. One of the young girls who I knew better than most asked if she could try, so I passed it to her while mentioning that, "The 'X' button erases what I

G. M. Bowen (✉)
Faculty of Education, Mount Saint Vincent University, 166 Bedford Highway, Halifax, NS B3M 2J6, Canada
e-mail: gmbowen@yahoo.com

E. Knoll
Département de Mathématiques, Université du Québec À Montréal, Case Postale 8888, Succ. Centre-Ville, Montréal, Québec H3C 3P8, Canada
e-mail: knoll.eva@uqam.ca

A. M. Willison
Independent Consultant, 6 Simcoe Place, Halifax, NS B3M 1H3, Canada
e-mail: willison.amy7@gmail.com

put in." She started playing with the Bee-Bot, trying different things. Other children shouted suggestions; some shouted reminders about the X button and what it did. Another child asked if they could try, so she passed it to them ... the playing continued. (G. M. Bowen)

STEM is an acronym that refers to curricula or practices that incorporate science, technology, engineering, and mathematics. The degree to which these different subjects need to be presented in a curriculum or activity for it to be considered STEM is widely discussed. For the purposes of this chapter, and consistent with others studying early learners (e.g., Milford & Tippett, 2016; Moomaw, 2013), we consider any curriculum/activity that intentionally emphasizes any two of the four disciplines to be STEM-oriented, including activities that use a technology (e.g., a robot) to accomplish learning in one of the other three disciplines. Within our early learning context, we use the term *preschool* to define the before-Kindergarten age group, generally from 3 to 5 years old. Some provinces provide formal preschool and others rely on either parental or daycare/childcare settings (Bowen et al., 2022).

STEM education for young learners can involve technological toys such as floor-based programmable robots (e.g., Bee-Bots, Botley® robots, Robot Mouse) that can be used with children under 5 years old. Many authors suggest that these floor robots could be introduced even before formal schooling begins (Lyons & Tredwell, 2015; McClure et al., 2017; McDonald & Howell, 2012). However, a recent review of Canadian provincial and territorial preschool curriculum/activity guidelines (Bowen et al., 2022) did not find any examples of such technologies.

Nevertheless, there is considerable room to incorporate technologies such as floor robots into early learning. Provincial government documents promote play-based approaches as being central to teaching and learning, which is consistent with research reports that young children learn best through play (Bodrova & Leong, 2015; Hewes, 2006; Russ & Wallace, 2013), and seems to apply to using floor robots such as the Bee-bot (Kwon et al., 2020; Vargová & Círus, 2021). In cases of specific learning outcomes, there often may be the need for various forms of instructor intervention such as direct instruction, guided play, Socratic questioning, or modeling of behaviours (Smith & Pellegrini, 2013).

Recent arguments have been made to include STEM in the early grades (Clements & Sarama, 2016). A joint position statement on technology and interactive media in early childhood states that "interactions with technology and media should be playful and support creativity, exploration, pretend play, active play, and outdoor activities" (National Association of Young Children & Fred Rogers Center for Early Learning & Children's Media, 2012, p. 7). Related to the need to support play, creativity, and exploration, Aronin and Floyd (2013), drawing from DeVries and Kohlberg (1987/1990), outlined four important principles to consider when engaging children with technology:

1. The student should be the source of the action to make the outcome more scientific.
2. The student should be able to see cause and effect relationships by changing the beginning action and seeing how it reflects the outcome.
3. The outcome of changing the variable must be observable to the preschooler.

4. The action and reaction must happen immediately for the child to see and make connections between cause and effect (p. 35).

Providing playful educational technology in the form of small programmable robots in an early learning setting may be one way to adhere to these principles and support meaningful science and STEM experiences for young children. Such experiences would also provide opportunities to learn coding—as called for by Papert (1993) and others—and to use robots such as the Bee-Bot (Campbell & Walsh, 2017; Komis & Misirli, 2016).

Although research on young learners' use of Bee-Bots has reported that they can be a catalyst for mathematical learning (De Michele et al., 2008; Highfield, 2010; Highfield et al., 2008) and that young children have demonstrated sustained engagement with the technology (Highfield, 2010; Magagna-McBee, 2010), there is little research on the use of Bee-Bots in the teaching and learning of science or STEM. In our review of academic literature on the Bee-Bot we noted that most of the research focused on mathematics outcomes with some studies of literacy, interpersonal communication, and programming (e.g., Beraza et al., 2010; Cacco & Moro, 2014; Leoste et al., 2022; McDonald & Howell, 2012; Schina et al., 2021) as well as computational thinking (Bhattacharya & Brown, 2020; Papadakis & Kalogiannakis, 2022). There was, however, little discussion of science or engineering in the studies—leaving the use of Bee-Bots to enhance understanding of those topics unexplored. Therefore, we collectively constructed a list of Bee-Bots' potential uses in early learning settings as shown in Table 9.1 (Bowen et al., 2022).

Table 9.1 Potential uses of Bee-Bots in early learning (adapted from Bowen et al., 2022)

Specific learning outcomes	Related subjects
Cause-and-effect	Science
Experimentation (e.g., friction/slopes)	Science
Directionality (e.g., left and right)	Mathematics, Literacy
Number lines	Mathematics
Number skills (e.g., number recognition, counting, arithmetic operations)	Mathematics
Graphing (e.g., histograms/bar charts)	Science, Mathematics
Letter/spelling skills (e.g., storytelling, letter identification, spelling)	Mathematics, Literacy
Shapes (e.g., squares, rectangles, circles)	Mathematics, Art
Distance/estimation	Mathematics
Strategies/games	Science, Mathematics
Algorithms/patterning	Mathematics, Coding
Vectors	Mathematics
Problem-solving	Science, Mathematics
Sequencing	Science, Mathematics, Literacy

In our own jurisdiction of Nova Scotia, there has been support for students in older grades to participate in robot design/build/task competitions for nearly 20 years (see Acadia Robotics, https://robots.acadiau.ca/about-us.html). In 2015, the provincial Department of Education and Early Child Development announced that the principles of coding would be included across the K–12 curricula (Casey, 2015). Following that announcement, the province provided various STEM technologies to every school with K–6 classes and offered workshops on how to incorporate those technologies into the curriculum (recently-introduced preschool programs for 4-year-olds in primary schools will also be able to access these technologies as appropriate). One of the provided technologies was the small programmable floor robot known as a Bee-Bot. Through the simple directional/movement button interface located on its back, the Bee-Bot allows early learners with easy access to programming. These buttons are considered to be embedded programming blocks, which are a form of visual programming (Yu & Roque, 2018) used with other technologies also in use in Nova Scotia schools (e.g., Scratch Jr, Scratch, Makey-Makey™, Sphero™, Sphero SPRK+™, Codey Rocky). Children in the early years who use Bee-Bots are, therefore, gaining experiences that prepare them for more advanced coding that is encountered in later grades.

Of course, teachers do not rely merely on ministerial or departmental workshops to find ideas on how to incorporate new technology into their classroom. Teachers look for credible information in other sources such as practitioner or academic journals and social media. A search of these resources is even more likely to occur with a technology such as Bee-Bots as early years educators "generally demonstrate a lack of knowledge and understanding about technology and engineering, and about developmentally appropriate pedagogical approaches to bring those disciplines into the classrooms" (Bers et al., 2013, p. 374). It is even more likely that a web media search was conducted by Nova Scotia instructors in the recently introduced pre-Kindergarten program since the program instructors were hired after the formal introductory workshops on the use of the new technologies were held. It is worth noting that in November, 2022 professional development provided by the local school board and focusing on incorporating technology in classrooms had only one session out of 40 that briefly discussed the Bee-bot. Additionally, Nova Scotia curriculum documents still do not mention their use in teaching subject matter, which emphasizes the importance of quality resources about Bee-bots being available.

Although there is research on the use of robots and the study of robotics in early learning settings (Bers et al., 2002; García-Valcárcel-Muñoz-Repiso & Caballero-González, 2019; Öztürk & Calingasan, 2019; Toh et al., 2016), most of that research focused on teaching and learning about robots and coding. There is little research, other than that previously cited, on learning other subjects (e.g., science, mathematics, and literacy) using robots and coding as part of the learning activity. Additionally, in most cases the existing research on using Bee-Bots in early learning settings is not published in mainstream educational journals but rather is found in proceedings papers or academic book chapters—which reduces the likelihood of knowledge transferring from the academic sphere to the early learning teaching practitioner on their effective use. Schrodt et al. (2021) is the sole exception we found. This issue is of particular interest for us because the oft-cited theory–practice gap (Guilfoyle et al., 2020; la Velle, 2019; McGarr et al., 2017) will never be bridged if practitioners are not able to find connections between proposed practices and academic research on those practices. This problem, we will note, is not unique to the use of the Bee-Bot (Bowen et al., 2020; Taylor et al., 2021). We attempted to collect all available online professional resources about using Bee-Bots in the classroom. We analyzed resources for (a) the type of information about Bee-Bots and recommendations for use with young children, (b) the subject/curriculum topics covered, and (c) the references provided to support the proposed or described practices using the Bee-Bot.

9.2 Methodology

Our interest in this project was to understand what online Bee-Bot resources were available to teachers and in what activities and subject areas teachers would be encouraged to use them. To compile the available resources, we engaged in two different but complementary approaches. First, we assumed that a teacher who was interested in using Bee-Bots would initially conduct an Internet search for resources on how they could be used in a classroom. Second, we compared the activities in the resources collected with the conceptual potential for Bee-Bots (Table 9.1) and evaluated their complexity using a scoring schema we developed (Table 9.3).

We hired a research assistant (RA) who was a 2nd year post-baccalaureate Bachelor of Education student to conduct the online search for Bee-Bot resources and to create a data set of the results. We provided the RA with an initial list of suggested search terms (i.e., Bee-Bot, Beebot, bee bot, curriculum, school, classroom), locations to explore, and guidelines for organizing the resources for analysis (e.g., PDF capture and creation of a data set). The first author encouraged the RA to search news media, manufacturer information, and retail sites because he had noted that students in his elementary STEM methods course often looked in these locations when investigating technologies like Bee-Bots. We asked the RA to search for resources that reflected what she would look for if she were intending to use Bee-Bots in her primary classroom. For example, our instructions regarding news media were: Find a collection of news articles that you would want to read if you were thinking about using Bee-Bots in your classroom.

The first type of resources that the RA searched for was Bee-Bot news articles using the Google News search website (http://news.google.com) and the search terms "Bee-Bot" and "Beebot;" 17 news media articles were located and entered into the data set. The second type of resources was YouTube videos that were located using the YouTube search engine and the same terms in the news media search; 430 videos that dealt with Bee-Bots were identified and entered in the data set. The third type of resources was lesson plans. Teacher planning websites were located using Google and the search phrase "teachers share lesson plans." At each website the RA searched for lesson plans using Bee-Bots and located nine teacher planning websites, only two of which had multiple Bee-Bot resources. Teachers Pay Teachers (TpT) and Share My Lesson (SML) each had 31 free resources on Bee-Bots. TpT had additional resources available for Bee-Bots at a cost, but only the 62 free resources from TpT and SML were added to the data set.

When the second author joined the team, she noticed that we had overlooked resources from Pinterest, a common source of teaching inspiration and resources for educators working in early learning settings (Navy & Nixon, 2021; Schroeder et al., 2019). We subsequently searched Pinterest using its search engine and two sets of search terms "bee bot activities primary" and "bee bot activities kindergarten." All English language Pins (i.e., links to either information stored in Pinterest or to another site) were saved into our primary and kindergarten boards (i.e., where pins are collated into collections within Pinterest). Over 200 Pins were saved in the primary board; upon review, we identified 186 Pins about Bee-Bots as opposed to bees or bots. We saved 200 Pins in the kindergarten board, which may have included Pins already saved in the primary board. Finally, using the Google search tool and the two search phrases (as above, but without quotation marks), the RA located five Bee-Bot retail/distributor sites and 13 teacher-created blog entries/webpages about Bee-Bots that discussed the use of Bee-Bots for early learning that were added to the data set.

9.2.1 Analytical Tools

To organize the analysis of these various Bee-Bot resources, we designed two code sheets. The first code sheet (Table 9.2) was used to assess the complexity of the TpT and SML lesson plans and the activities they described. The second code sheet (Table 9.3) allowed us to assess the focus of each resource (i.e., description or teaching plan), the number and type of activities, the type of prompting recommended, and other aspects such as suggested number of children, materials required, and references cited. The authors designed the code sheets based on their experiences working with children and Bee-Bots and from a previous examination of pre-primary curriculum documents (Bowen et al., 2022). The code sheets were piloted and modified by adding new categories that were needed to adequately code resources.

Here is an example of how scoring using the first code sheet would occur. A lesson plan activity describing the use of a Bee-Bot has a topic (1 point), a task to be accomplished (0.5, 1.0, or 1.5 points), a description of the Bee-Bot moving forward (1 point), and a return to the start (1 point). Thus, the base score for any Bee-Bot lesson plan in the data set would be 3.5 points. Complexity scores would be obtained by summing the checkmarks (1 point each) in the row and adding the base score.

The first author analyzed resources using the second code sheet then coded and established new codes or categories either individually or through consultation with another author. The coding for each resource item was entered into a spreadsheet.

Table 9.2 Code sheet for activity complexity of TpT and SML plans

Title/Code: _____

Teaching Supports

□ Demo	□ Lesson Plan	□ Task Cards	□ Grid ___ x ___	
□ Floor/Maze/Route/Map		□ Resources <u>only</u>	□ Game (vs. task)	□ Programming Cards
□ Planning Route Sheet		□ Resources to Print		

Topic[a]	Task Given / Scaffolded / Created[b] 0.5 / 1 / 1.5	Forward[c]	Return[d]	Simple Sequence[e]	Jump Sequence[f]	Complexity[g]	End Action[i]	Finish Point[j]	Total
Letters □	G / S / C	□	□	□	□	Spell Word □	□	□	
Colours □	G / S / C	□	□	□	□	?	□	□	
Shapes □	G / S / C	□	□	□	□	?	□	□	
Numbers □	G / S / C	□	□	□	□	Calculation □	□	□	
Stories □	G / S / C	□	□	2-part story[h] □	Long story □	Spontaneous □	□	□	
Images □	G / S / C	□	□	□	□	Combination □	□	□	
				Pairing[k] □	Multi-pairing[l] □	Sequence □			

[a]Was the Bee-Bot route based on letters, colours, shapes, numbers, stories, or images?
[b]Were the students given a route to follow, scaffolded toward creating a route, or creating the route themselves?
[c]Was the route forward from the start to a particular target?
[d]Did the Bee-Bot return from the target to the start?
[e]Did the Bee-Bot enact a simple ordered sequence? (e.g., spell "cat"; go to a number and then the next number)
[f]Did the Bee-Bot follow a jump sequence? (e.g., go to the letter A, then go to two letters after A)
[g]What was the task complexity? (e.g., spelling a word, completing a calculation, creating a spontaneous story, creating a combination or sequence of images [e.g., types of food, predators])
[h]Does the Bee-Bot enact a short story or a long story? Is it a spontaneous or planned story?
[i]Does the Bee-Bot perform a programmed end action? (e.g., spin)
[j]Is there a point that indicates the task is finished or is it free play without a defined end-point?
[k]Is it a pairing activity (e.g., two flowers, predator/prey, happy/sad faces)
[l]Is it a multi-pairing activity? (e.g., go to two dogs, and then two cats)

The appropriate codes or text descriptions were entered into each associated cell according to the instructions on the code sheet.

9.3 Analysis, Results, and Discussion

Our collection of available resources resulted in a dataset that represented a snapshot of what was available in July and August of 2020, when the search was conducted. Given the range of resource types that were located, we adopted a variety of analytic and summarization approaches when examining those resources. Using the code sheets and enumerated information in those situations where we believed details would provide useful insights about a particular resource type, we relied on developing qualitative and relational descriptors (e.g., many, few) and listing the variety of perspectives available. This approach recognized that our review of available resources was far more thorough than an average teacher would engage in when exploring the possibilities of Bee-Bots for their classroom.

Table 9.3 Code sheet for analysis of Bee-Bot resources (Pinterest, Teacher Blogs, and Websites)

Title/Code: _____

(A) Basic context		
1	Description	Description of the item
2	Type of Source	Pinterest, teacher blog, teacher website, etc
3a	Orig Source	URL link
3b	Refer to:	If #2 refers somewhere, use same list as #2. For instance, Pinterest may refer to TpT
4a	Content	1 = intro, 2 = activity instructions, 3 = both, 4 = refers to/describes activity, 5 = other, 6 = n/a
4b	# Activities	# of Activities <u>described</u> in resource (vs. <u>mentioned</u>), may be kids "doing it" vs. Lesson Plan
(B) Prompting type: Score 1 = Yes, 0 = No, 2 = n/a, 3 = not specified		
4c	Direct Instruct	Does the adult give direct instructions to the students?
5	Guide/ Suggest	Does the adult make suggestions to guide the students?
6	SocraticQ	Does the adult ask questions/set up problems to implicitly guide the students?
7	Model	Does the adult model activities/actions with the robot?
8a	Explore	Open exploration of Bee-Bot by students
8b	Peer	Are students supposed to seek guidance from other students? OR is it a group activity?
9	Other	
(C) Type of activity: Score 1 = Yes, 0 = No, 2 = n/a | What instructions are in the document? If not an instruction document, then 2		
10	Game	Are the students engaging in a game?
11	Spell	Are the students spelling something (on a mat)?
12	Story	Are the students being encouraged to tell a story?
13a	Mathematics	Are the students engaging in mathematics activities? (incl. shapes)
13b	Algorithms	Are the students using/learning algorithms (vs. using algorithms)
14	Direction	Are the students determining directions? (i.e., left/right) (vs. using directions)
15	Program	Are the students learning to program? Are their specific ties? (incl. laying out cards)
16	Route	Are the students learning to follow a route?
17	Problem Solve	Are the students learning to problem solve?
18	Maps Created	Are the students creating a map or mat?
19	Other	Describe textually …
(D) Other		
20	Individ/Grp	Individ = 1, Group = 2, Individ w other individ = 3, Group w other groups = 4, 5 = n/a, 6 = not specified [Note: 3 & 4 are for interactions between those entities]

(continued)

Table 9.3 (continued)

Title/Code:		
21	#Bee-Bots	How many Bee-Bots are used in the activity <u>by each student/group</u>?
22	Costume?	Are students making costumes for the Bee-Bot? 0 = no, 1 = yes, 0 = no, 1 = yes, 2 = n/a
23	Intro/Basic/Adv	Does the activity seem <u>I</u>ntroductory? <u>B</u>asic? <u>A</u>dvanced? (I/B/A)
(E) Are resources available: Score 1 = Yes, 0 = No, 2 = n/a | Are there resources in the document? If not an instruction document, then 2		
24	Cards	Are there cards provided for the students?
25	Maps Provided	Are the students using a map/mat?
26	Other Resources	Describe in text
27	Subj/Topic	Describe in text
(F) References: Does the article provide references or supports?		
28	Acad B/J	Reference to academic books or journals 0 = none, 1 = journals, 2 = books, 3 = both
29	Practitioner	Reference to practitioner books or journals 0 = none, 1 = journals, 2 = books, 3 = both
30	Mag	Magazine article(s) 0 = no, 1 = yes
31	News	Newspaper article(s) 0 = no, 1 = yes
32	Textbooks	Textbook(s) 0 = no, 1 = yes
33	Website	Website(s) 0 = no, 1 = yes
34	Pics/Vids	Bee-Bot shown <u>IN USE</u> in 1 = video, 2 = pictures, 3 = both, 4 = n/a, 0 = none

9.3.1 News Media

The news media articles were independently read by two of the authors, who each took notes regarding main themes. Discussions following reading established consensus on main themes and led to the claims that follow. Most of the 17 news media articles from the data set described classroom use of Bee-Bots and included interviews with students and their teachers. Although Bee-Bots can be used with very young children, 12 articles focused on the necessity of preparing students for technology-oriented careers or jobs especially with respect to *coding* or *programming*—terms that were used 144 times (determined by copying the text to a word processor and using the word-count feature). Many articles discussed Bee-Bots as developing foundational skills (e.g., communication, problem solving, groupwork) and being used in topics such as mathematics, geography, and literacy. One article discussed the role that Bee-Bots could play when working with autistic students. Finally, given the experience of the first author, we were surprised that the terms *fun, enjoyment,* and *play* were used only 22 times in total across all 17 articles.

Table 9.4 Subjects and topics demonstrated in Bee-Bot YouTube videos

Coding and Bee-Bot focus	Content area focus

• Programming scratch
• How to use vinyl grid sheets with Bee-Bots, how to make cards
• Children using planning grids, cards
• Using Bee-Bot emulator & app
• Demonstrating grids/mazes/obstacle courses made with:
  ◦ Centicubes
  ◦ Straws
  ◦ Lego®
  ◦ Pencils
  ◦ Chalk
  ◦ Whiteboards
  ◦ Cardboard
	• Directionality (e.g., with mazes)
	• Literacy
	  ◦ Phonics
	  ◦ Storytelling/readalouds
	  ◦ Vocabulary
	  ◦ Spelling
	• Science
	• Geography
	• Mathematics
	  ◦ Number lines
	  ◦ Coding
	  ◦ Sequencing
	  ◦ Counting
	  ◦ Skip counting
	  ◦ Patterns, repetition
	• Fine Arts
	  ◦ Drawing
	  ◦ Dance & choreography of Bee-Bots
	  ◦ Making/using costumes

9.3.2 YouTube Videos

A search of YouTube located 183 English language videos that specifically dealt with Bee-Bots; another 247 videos were available in other languages including French, Spanish, Romanian, Czechoslovakian, Dutch, Finnish, Estonian, Portuguese, Russian, Bulgarian, German, and Greek. We watched all of the English language videos that were less than 5 min long in their entirety; most of the videos fell into this category. We viewed longer videos for 2 min at the beginning and then skipped through until the end, viewing in more detail if anything relevant was noted in the skip. We noted the topics explored, approaches taken, and if children were part of the video. Of the 166 English language videos that showed children using the Bee-Bot in a variety of subject areas (Table 9.4), we could not always discern how old the children were although they could be described as early learners; 17 videos did not show children. We did not formally code the videos for complexity; however, our viewing notes suggested that there were few examples of complex or higher-order Bee-Bot use depicted.

9.3.3 Lesson Planning Websites

Of the 62 resources available from the two lesson planning websites, we randomly chose 10 TpT and 10 SML lessons to analyze for complexity using the code sheet

(Table 9.2). We distinguished between lesson plans describing the use of Bee-Bots as a central part of the lesson versus plans mentioning Bee-Bots as something that might be used in a lesson, referring to them as *primary use* and *secondary use*, respectively. Further, we distinguished between these two categories and plans that described resources available to be used with Bee-Bots but that had no suggested application or curricular outcome, labeling them as *tertiary use*. We noted that many of the lesson plan descriptions had associated resources to print. The following are examples of what was considered acceptable text to describe a primary use as opposed to a secondary use of Bee-Bots.

<u>Examples of primary use:</u>

"Children have the Bee-Bot and one map. They say where they would like to get the Bee-Bot to, and then program him to go to their chosen destination. Did they manage to get him to the right place? Why/why not? They must remember to press clear after every turn."

"Children have the large shape cards and the Bee-Bot. They challenge each other to get him to the correct shape."

<u>Examples of secondary use:</u>

"Working with the Bee-Bots to make left and right turns."

"Bee-Bot?"

"Use cvc Bee-Bot mat. Sound talk a word – children swat [*sic*] it."

We located 31 lesson plans from each of the TpT and SML websites and archived these resources as PDF files. We evaluated lesson plan content as either primary, secondary, or tertiary use for teachers. TpT had 18 primary and 13 tertiary use lesson plans; SML had 13 primary, 13 secondary, and 5 tertiary use lesson plans. Using the complexity code sheet (Table 9.3), we analyzed 10 randomly chosen, primary use Bee-Bot lesson plans from each of TpT and SML. Results suggested that the majority of the free Bee-Bot lesson plans from these two sites have low complexity scores as shown in Fig. 9.1. Our coding indicated that few free lesson plans involved student choice (e.g., using the Bee-Bot to tell stories). Regarding topic, nine lesson plans were mathematics oriented (e.g., counting, calculation, use of coordinates), four were literacy oriented (e.g., spelling, stories, word recognition), and only one was science oriented (e.g., sounds).

9.3.4 Pinterest Resources

To better understand what Bee-Bot resources were available in Pinterest, we randomly opened 30 Pins from the 186 that were entered into our data set. Creating PDF versions of the actual Pins turned out to be difficult—a challenge that we concluded was an intentional feature of Pinterest, so those resources were not archived but instead were analyzed in situ using the complexity code sheet (Table 9.3). We also noted if the Pin was linked to an external site, such as TpT. There were 16 Pins with direct links to paid TpT resources, 10 Pins linked to online stores with

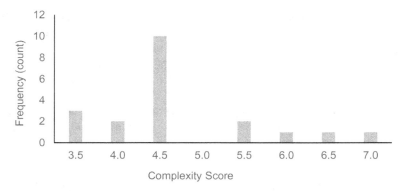

Fig. 9.1 Complexity of Bee-Bot lesson/activity plans available

resources such as Bee-Bot Jackets or grid cards or to teacher websites with Bee-Bot related printables, two Pins linked to YouTube videos, one Pin linked to a picture of a Bee-Bot, and one Pin linked to information at a Google account. We did not find any lesson plans or described activities linked to the 30 Pins we analyzed. Overall, about two-thirds of the Pins were linked to sites selling curriculum documents or resources.

9.3.5 Bee-Bot Retail Sites

Of the five retail sites identified by the RA, four sites merely presented the Bee-Bot and associated products with descriptions (e.g., with respect to functionality), affective comments (e.g., exciting, new), anthropomorphic statements (e.g., friendly, cute), and general comments about classroom utility (e.g., good for teaching sequences). The fifth site was the main distributor's website that included a list of available products and provided Bee-Bot resources including a blog page for teachers. When the Bee-Bot was released 15 years ago, a reasonably detailed resource guide was provided; however, it is no longer available at the distributor's website. The RA did not find this document in her searches, which suggests that many teachers might not readily find it either.

9.3.6 Teacher Blog Entries and Webpages

It is not uncommon for teachers to post information about teaching resources or classroom activities in either blog entries or webpages (henceforth referred to as teacher web resources). We analyzed the 13 teacher web resources in our data set using the second code sheet (Table 9.3); seven had descriptions of Bee-Bot-based activities for early learners, 11 had introductory information about Bee-Bots, while two had no introductory text and only referred to (but did not describe) activities. Of the five resources that provided details for conducting an activity with Bee-Bots, all used Socratic questioning techniques; only two suggested using open-play approaches.

The types of activities found in the teacher web resources were often not curriculum oriented (i.e., not related to a typical subject or topic in early learner curriculum). Only three resources described developing understanding of a science, mathematics, or literacy topic; seven activities described in these resources taught concepts of programming, three used command cards (Fig. 9.2), and four used a map or route diagram. All of the activities used only one Bee-Bot per student or group, and all corresponded with a complexity score from 3.5 to 4.5 as per Table 9.2. Seven of the 13 teacher web resources included images and/or videos of a Bee-Bot in use. None of the teacher web resources included references to academic books or journals, practitioner articles, or other sources such as magazines, newspapers, or textbooks.

9.4 Observations and Implications

From the Bee-Bot resources located for this study, we identified four trends: (a) a narrow curricular emphasis, (b) a lack of complexity, (c) a lack of materials to scaffold programming, and (d) a limited reference to play-based based approaches. We also noted a dearth of empirical evidence of learning with Bee-Bots and of published peer-reviewed curricular materials; the sole exception was that of Schrodt et al. (2021).

First, we noted that many of the curriculum uses in Table 9.1 (e.g., cause-and-effect, graphing, problem solving) appeared infrequently or were non-existent in the resources we examined. Science was an infrequent topic in Bee-Bot resources. Although fundamental topics in mathematics (e.g., counting, number recognition) and literacy (e.g., spelling, letter recognition, word meaning) were popular, higher order aspects of those topics (e.g., skip-counting, word pairing, story telling) were less frequently presented.

Second, we noted a lack of Bee-Bot resources that would help teachers incorporate activities of a higher lever of complexity as scored using the code sheet (Table 9.2). This lack of resources for more complex activities places a considerable responsibility on teachers. Although there are multiple ways to use the Bee-Bot in an early learning environment (as per Table 9.1), even the distributor does not provide resources for using Bee-Bots in more complex ways to develop higher-order thinking skills.

Third, we noticed a lack of materials to support scaffolding of programming, such as command cards or programming guides (see examples in Fig. 9.2). These scaffolding materials can support students in planning Bee-Bot routes and moves as an intermediate step between following a prescribed task and mentally planning and programming a route.

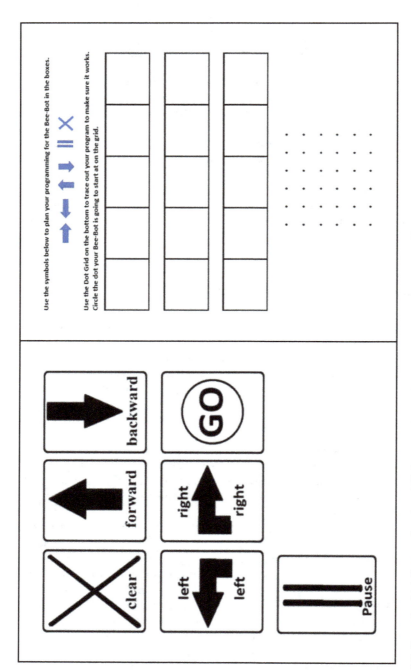

Fig. 9.2 Example of (a) Command cards and (b) Programming guide

Fourth, we noted limited mention of play-based approaches to using the Bee-Bot. Most resources that we found involved explicit instruction; few resources involved any aspect of free or guided play despite the importance of play-based approaches in early learning (Bowen et al., 2022; Guirguis, 2018; Lifter et al., 2011; Lillard et al., 2013). A focus on explicit instruction was particularly noticeable in newspaper articles; this seemed to imply that the role of education, even for early learners, is to support a neoliberal goal of a globally competitive economy (Adams, 2013; Kelly, 2013).

Finally, to date there is little published research on incorporating Bee-Bots with early learners in science and STEM. To address this lack, our future plans involve conducting research at an early learning centre to examine (a) what learning occurs when young children engage with Bee-Bots in a free-play situation and (b) how using Bee-Bots can influence learning in guided-play situations, particularly in the context of science.

Overall, we conclude that Bee-Bot use with early learners is under-conceptualized and that the current readily available resources are not well informed by research. We note that using Bee-Bots with early learners often requires children's careful attention to planning; yet such planning of Bee-Bot movement during the activities (e.g., using scaffolds as shown in Table 9.2) was not emphasized in many of the resources that we located. Additionally, many of these resources involved more teacher instruction than the guided involvement suggested for effective play-based approaches (Smith & Pellegrini, 2013). The Bee-Bot can introduce children to coding, robots, and problem solving while supporting science, mathematics, and literacy exploration; but few of the resources that we located appeared to do that well. Bee-Bots can clearly be used to support STEM education through engagement with technology and problem-solving while supporting the development of science and mathematics concepts. However, from our review of available materials, we have concluded that there is a need for better quality Bee-Bot resources for early learning classroom instructors. At present, it seems that high-quality educator resources are scarce, particularly in the social media formats that we explored.

References

Adams, J. (2013). The artful dodger: Creative resistance to neoliberalism in education. *Review of Education, Pedagogy, & Cultural Studies, 35*(4), 242–255.

Aronin, S., & Floyd, K. K. (2013). Using an iPad in inclusive preschool classrooms to introduce STEM concepts. *Teaching Exceptional Children, 45*(4), 34–39.

Beraza, I., Pina, A., & Demo, B. (2010). Soft & hard ideas to improve interaction with robots for kids & teachers. *SIMPAR 2010 Workshop Proceedings, 2nd International Conference on Simulation, Modeling, and Programming for Autonomous Robots* (pp. 549–555). https://www.terecop.eu/SIMPAR2010/TR-TWR-2010/19-TeachingRobotics.pdf

Bers, M. U., Ponte, I., Juelich, C., Viera, A., & Schenker, J. (2002). Teachers as designers: Integrating robotics in early childhood education. *Information Technology in Childhood Education Annual, 1*, 123–145.

Bers, M. U., Seddighin, S., & Sullivan, A. (2013). Ready for robotics: Bringing together the T and E of STEM in early childhood teacher education. *Journal of Technology & Teacher Education, 21*(3), 355–377. https://sites.tufts.edu/devtech/files/2018/02/BringingTogetherT.pdf

Bhattacharya, P., & Brown, M. (2020). Bee-bot for computational thinking: An artifact analysis. In D. Schmidt-Crawford (Ed.), *Proceedings of Society for Information Technology & Teacher Education International Conference* (pp. 2–7). Association for the Advancement of Computing in Education.

Bodrova, E., & Leong, D. J. (2015). Vygotskian and post-Vygotskian views on children's play. *American Journal of Play, 7*(3), 371–388. https://www.museumofplay.org/app/uploads/2022/01/7-3-article-vygotskian-and-post-vygotskian-views.pdf

Bowen, G. M., Knoll, E., & Willison, A. M. (2022). Bee-Bot robots and their STEM learning potential in the play-based behaviour of preschool children in Canada. In S. D. Tunnicliffe & T. J. Kennedy (Eds.), *Play and STEM Education in the early years: International policies and practices* (pp. 181–198). Springer.

Bowen, G. M., Taylor, J., Patrick, P., Summers, R., Kubsch, M., Warfa, A., Sezen-Barrie, A., Guzey, S., & LaChapelle, C. (2020). Understanding the use of academic research in science education practitioner journals. *Journal of the Canadian Association of Curriculum Studies, 18*(1), 151–153. https://jcacs.journals.yorku.ca/index.php/jcacs/article/view/40452

Cacco, L., & Moro, M. (2014). When a Bee meets a sunflower. *Proceedings of 4th International Workshop Teaching Robotics Teaching with Robotics & 5th International Conference on Robotics in Education* (pp. 68–75). https://www.terecop.eu/TRTWR-RIE2014/files/00_WFr1/00_WFr1_09.pdf

Campbell, C., & Walsh, C. (2017). Introducing the 'new' digital literacy of coding in the early years. *Practical Literacy: The Early and Primary Years, 22*(3), 10–12. https://www.jcu.edu.au/__data/assets/pdf_file/0006/526191/Campbell-and-Walsh-2017.-Coding-as-the-new-digital-literacy-.pdf

Casey, K. (2015). *Minister announces coding as a priority during education day.* https://novascotia.ca/news/release/?id=20151021002

Clements, D. H., & Sarama, J. (2016). Math, science, and technology in the early grades. *The Future of Children, 26*(2), 75–94. http://www.jstor.org/stable/43940582

De Michele, M. S., Demo, G. B., & Siega, S. (2008). A Piedmont SchoolNet for a K-12 minirobots programming project: Experiences in primary schools. *SIMPAR 2008 Workshop Proceedings, International Conference on Simulation, Modeling and Programming for Autonomous Robots* (pp. 90–99). https://www.terecop.eu/downloads/simbar2008/MicheleDemoSiega.pdf

DeVries, R., & Kohlberg, L. (1987/1990). *Constructivist early education: Overview and comparison with other programs.* National Association for the Education of Young Children.

García-Valcárcel-Muñoz-Repiso, A., & Caballero-González, Y. A. (2019). Robotics to develop computational thinking in early childhood education. *Comunicar, 59*, 63–72. https://doi.org/10.3916/C59-2019-06

Guilfoyle, L., McCormack, O., & Erduran, S. (2020). The "tipping point" for educational research: The role of pre-service science teachers' epistemic beliefs in evaluating the professional utility of educational research. *Teaching & Teacher Education, 90*, Article 103033. https://doi.org/10.1016/j.tate.2020.103033

Guirguis, R. (2018). Should we let them play? Three key benefits of play to improve early childhood programs. *International Journal of Education & Practice, 6*(1), 43–49. https://doi.org/10.18488/journal.61.2018.61.43.49

Hewes, J. (2006). Let the children play: Nature's answer to early learning. *Canadian Council on Learning/Lessons in Learning.* https://galileo.org/earlylearning/articles/let-the-children-play-hewes.pdf

Highfield, K. (2010). Robotic toys as a catalyst for mathematical problem solving. *Australian Primary Mathematics Classroom, 15*(2), 22–27. https://files.eric.ed.gov/fulltext/EJ891802.pdf

Highfield, K., Mulligan, J., & Hedberg, J. (2008). Early mathematics learning through exploration with programmable toys. In O. Figueras, J. L. Cortina, S. Alatorre, T. Rojano, & A.

Sepúlveda (Eds.), *Proceedings of the Joint Meeting of PME 32 and PME-NA XXX* (Vol. 3, pp. 169–176). http://www.pmena.org/pmenaproceedings/PMENA%2030%202008%20Proceed ings%20Vol%203.pdf

Kelly, C. (2013). Primary education: From market forces to personal development. In M. Allen & P. Ainley (Eds.), *Education beyond the coalition: Reclaiming the agenda* (pp. 24–47). Radicaled. https://radicaledbks.files.wordpress.com/2013/09/clare-kelly-primary.pdf

Komis, V., & Misirili, A. (2016). The environments of educational robotics in early childhood education: Towards a didactical analysis. *Educational Journal of the University of Patras UNESCO Chair, 3*(2), 238–246.

Kwon, U. J., Nam, K. W., & Lee, J. H. (2020). Exploring the effects of unplugged play for children aged 3, 4 and 5-based on Bee-bot. *International Journal of Advanced Culture Technology, 8*(2), 239–245.

la Velle, L. (2019). The theory–practice nexus in teacher education: New evidence for effective approaches. *Journal of Education for Teaching, 45*(4), 369–372. https://doi.org/10.1080/026 07476.2019.1639267

Leoste, J., Tammemäe, T., Eskla, G., San Martín López, J., Pastor, L., & Blasco, E. P. (2022). Bee-bot educational robot as a means of developing social skills among children with autism-spectrum disorders. In M. Merdan, W. Lepuschitz, G. Koppensteiner, R. Balogh, & D. Obdržálek (Eds.). *Robotics in education* (pp. 14–25). Springer. https://doi.org/10.1007/978-3-030-82544-7_2

Lifter, K., Foster-Sanda, S., Arzamarski, C., Briesch, J., & McClure, E. (2011). Overview of play: Its uses and importance in early intervention/early childhood special education. *Infants & Young Children, 24*(3), 225–245. https://doi.org/10.1097/IYC.0b013e31821e995c

Lillard, A. S., Lerner, M. D., Hopkins, E. J., Dore, R. A., Smith, E. D., & Palmquist, C. M. (2013). The impact of pretend play on children's development: A review of the evidence. *Psychological Bulletin, 139*(1), 1–34. https://doi.org/10.1037/a0029321

Lyons, C. D., & Tredwell, C. T. (2015). Steps to implementing technology in inclusive early childhood programs. *Computers in the Schools, 32*(2), 152–166. https://doi.org/10.1080/07380569. 2015.1038976

Magagna-McBee, C. A. (2010). *The use of handheld devices for improved phonemic awareness in a traditional kindergarten classroom* [Unpublished doctoral dissertation]. Walden University.

McClure, E. R., Guernsey, L., Clements, D. H., Bales, S. N., Nichols, J., Kendall-Taylor, N., & Levine, M. H. (2017). *STEM starts early: Grounding science, technology, engineering, and math education in early childhood.* Joan Ganz Cooney Center at Sesame Workshop. https://joanganzc ooneycenter.org/wp-content/uploads/2017/01/jgcc_stemstartsearly_final.pdf

McDonald, S., & Howell, J. (2012). Watching, creating and achieving: Creative technologies as a conduit for learning in the early years. *British Journal of Educational Technology, 43*(4), 641–651. https://doi.org/10.1111/j.1467-8535.2011.01231.x

McGarr, O., O'Grady, E., & Guilfoyle, L. (2017). Exploring the theory–practice gap in initial teacher education: Moving beyond questions of relevance to issues of power and authority. *Journal of Education for Teaching, 43*(1), 48–60. https://doi.org/10.1080/02607476.2017.1256040

Milford, T. M., & Tippett, C. D. (2016). The design and validation of an early childhood STEM classroom observational protocol. *International Research in Early Childhood Education, 6*(1), 24–37. https://doi.org/10.4225/03/5817cdcd6b1e8

Moomaw, S. (2013). *Teaching STEM in the early years: Activities for integrating science, technology, engineering, and mathematics.* Redleaf Press.

National Association for the Education of Young Children & Fred Rogers Center for Early Learning & Children's Media at Saint Vincent College. (2012). *Technology and interactive media as tools in early childhood programs serving children from birth through age 8* [Position statement]. https://www.naeyc.org/sites/default/files/globally-shared/downloads/PDFs/resources/pos ition-statements/ps_technology.pdf

Navy, S. L., & Nixon, R. S. (2021, April 8–12). *Beyond cute: Examining the quality of Pinterest as a resource for science teachers* [Paper presentation]. American Educational Research Association Virtual Annual Meeting.

Öztürk, H. T., & Calingasan, L. (2019). Robotics in early childhood education: A case study for the best practices. In Information Resources Management Association (Ed.), *Early childhood development: Concepts, methodologies, tools, and applications* (pp. 892–910). IGI Global. https://doi. org/10.4018/978-1-5225-7507-8.ch044

Papadakis, S., & Kalogiannakis, M. (2022). Learning computational thinking development in young children with Bee-Bot educational robotics. In M. Kalogiannakis & S. Papadakis (Eds.). *Handbook of research on tools for teaching computational thinking in P-12 education* (pp. 926–947). IGI Global. https://doi.org/10.4018/978-1-7998-4576-8.ch011

Papert, S. (1993). *Mindstorms: Children, computers, and powerful ideas* (2nd ed.). Basic Books.

Russ, S. W., & Wallace, C. E. (2013). Pretend play and creative processes. *American Journal of Play, 6*(1), 136–148. https://www.museumofplay.org/app/uploads/2022/01/6-1-article-pretend-play.pdf

Schina, D., Esteve-Gonzalez, V., & Usart, M. (2021). Teachers' perceptions of Bee-bot robotic toy and their ability to integrate it in their teaching. In W. Lepuschitz, M. Merdan, G. Koppensteiner, R. Balogh, & D. Obdržálek (Eds.). *Robotics in education: Advances in intelligent systems and computing.* Springer. https://doi.org/10.1007/978-3-030-67411-3_12

Schrodt, K., Winters, J., & Huddleston, T. (2021). A STEAM_ed bee. *Science & Children, 58*(5), 86–90. https://www.nsta.org/science-and-children/science-and-children-mayjune-2021/steamed-bee

Schroeder, S., Curcio, R., & Lundgren, L. (2019). Expanding the learning network: How teachers use Pinterest. *Journal of Research on Technology in Education, 51*(2), 166–186. https://doi.org/10.1080/15391523.2019.1573354

Smith, P. K., & Pellegrini, A. (2013). Learning through play. In R. E. Tremblay, M. Boivin, & R. DeV. Peters (Eds.), *Encyclopedia on early childhood development* [online]. https://www.child-encyclopedia.com/play/according-experts/learning-through-play

Taylor, J. A., Bowen, G. M., Kubsch, M., Summers, R., Patrick, P., Warfa, A., Lachapelle, C., Sezen-Barrie, A., & Guzey, S. (2021, April 7–10). *Translating research into classroom practice: Who is publishing in science education practitioner journals (SEPJs)?* [Paper presentation]. National Association for Research in Science Teaching Virtual 94th Annual International Conference.

Toh, L. P. E., Causo, A., Tzuo, P.-W., Chen, I.-M., & Yeo, S. H. (2016). A review on the use of robots in education and young children [Special issue]. *Educational Technology & Society, 19*(2), 148–163. http://www.jstor.org/stable/jeductechsoci.19.2.148

Vargová, M., & Círus, L. (2021). The use of a Bee-bot in pre-primary and primary education. *Journal of Education, Technology and Computer Science, 32*(2), 45–50.

Yu, J., & Roque, R. (2018, June). A survey of computational kits for young children. *Proceedings of the 17th ACM Conference on Interaction Design and Children* (pp. 289–299). https://doi.org/10.1145/3202185

G. Michael Bowen BSc (University of Guelph), BJ (University of King's College), BEd (Dalhousie University), MSc (University of Guelph), MA (University of Guelph), PhD (University of Victoria) is an associate professor of science and STEM education in the Faculty of Education at Mount Saint Vincent University in Halifax, Nova Scotia. He has taught science in public and private schools in Canada and the United States, and in urban and rural settings from Grades 5 to 13. His university teaching includes science education, STEM education, research methods, curriculum theory, identity theory, critical media literacy, environmental education, and action research. His past research has examined graphical and data literacy, inquiry science, informal science education, online technologies, news media and entertainment media influences on the public understanding of science and scientists, preservice science teacher preparation, and science fairs. His current research interests focus on the use of STEM technologies with young children, university student knowledge calibration during assessment, and the development of student understanding of science inquiry using inexpensive, homemade probe technologies.

Eva Knoll BSc (Architecture, McGill), MSc (Aménagement, Université de Montréal), PhD (Education, Exeter University), Visual Arts Certificate (Textiles, NSCAD), is an associate professor of mathematics education in the Département de mathématiques of the Université du Québec à Montréal. Using a research-creation approach linked to her own art practice, she investigates the mathematical reasoning and learning involved in processes and experiences of art creation. In addition to papers, presentations and workshops, this research produces collections of artefacts that embody her findings.

Amy Willison BA (University of Waterloo), is a certified Early Childhood Educator with 25 years experience working with both preschool and school-age children in a wide variety of settings. Her work with children focuses on using emergent play-based approaches and interactive story-telling. She has both taught ECE students and supervised them on practicum. During the pandemic she returned to post-secondary education to obtain qualifications that will enable her to work with literacy and reading programs with young children in library settings.

Chapter 10
Connecting Children to Nature Through Scientific Inquiry: The Impact on Children's Well-Being

Michael W. Link

10.1 Introduction

What is the purpose of education? It might be argued that education should provide opportunities for students to live well within the constraints of a finite planet. But what is needed to live well? Research in this field points to many things, such as social connections, a feeling of security, and time for leisure and creative pursuits (Cacioppo et al., 2008; Gordon & O'Toole, 2015; Mannell, 2007; Seppala et al., 2013). However, two of the most important factors in living well are agency and purpose (e.g., Kern et al., 2015; Mcknight & Kashdan, 2009; Nussbaum, 2013; Seligman, 2011). That is, do we have the ability to choose to take on pursuits we find meaningful? The practice of science is at its heart purpose-driven. In science, we explore questions that we have about the natural world and natural phenomena; in engineering, we apply the principles of science to try and solve problems. We also need particular capabilities and a sense of agency to explore questions and solve problems that we find meaningful. The facilitation of science inquiry may, therefore, be an ideal domain to nurture student well-being. Further, when the practice of science takes us out of the classroom or laboratory and into nature, students can experience the benefits of immersion in the natural world.

M. W. Link (✉)
University of Winnipeg, Winnipeg, MB, Canada
e-mail: m.link@uwinnipeg.ca

© The Author(s), under exclusive license to Springer Nature Switzerland AG 2023   167
C. D. Tippett and T. M. Milford (eds.), *Exploring Elementary Science Teaching and Learning in Canada*, Contemporary Trends and Issues in Science Education 53,
https://doi.org/10.1007/978-3-031-23936-6_10

10.2   The Importance of a Connection to Nature

Children, given their seemingly innate capacity for curiosity and wonder, have been described by Piaget as "little scientists," engaged explorers seeking to understand the world around them (as cited in Bee & Boyd, 2010, p. 144). However, instead of spending their days involved in activities such as making mud pies, chasing butterflies, and building forest dens, children are more likely to be sitting indoors and engaged with technology. On average, youth spend over 49 h per week viewing media, with the vast majority having access to a bedroom television, video-game console, and a mobile phone (Strasburger et al., 2010). The COVID-19 pandemic saw an increase in the amount of time children and youth spend in front of screens. One survey of children from Ontario indicated that screen time not associated with online schooling increased by over 3 h per day during the COVID-19 pandemic (Seguin et al., 2021). Research has since explored the mostly negative and interconnected impacts of increased screen time on movement behaviours, sleep, and mental health (e.g., Kharel et al., 2022; Lien, 2022; Vézina-Im et al., 2022).

While children spent over 7 h per day in front of screens, they are reported to be spending only 4–7 min per day in unstructured play outdoors. Time spent apart from nature is a substantial change from previous generations, compounded by the fact that many children are multitasking their viewing, meaning they may be looking at their phone at the same time as they are watching television. Time spent in front of screens comes at the expense of time spent outdoors. A survey in the United Kingdom found that three-quarters of children spent less time outside than prison inmates (Carrington, 2016).

As people spend more of their childhood indoors, they grow increasingly disconnected from the natural world (Louv, 2011). Ecopsychologists contend that this has a detrimental effect on the well-being of children and the health of the planet (Gupta et al., 2018; Kahn & Hasbach, 2012). Much research has focussed on the benefits of time spent outdoors to human health and well-being. Many studies have shown that nature contact provides a range of physical, cognitive, and social/emotional benefits. For example, researchers of a Toronto-based study investigating the relationship between urban trees and human well-being found that residents living in neighbourhoods with many trees, particularly large trees, described feeling healthier than residents in neighbourhoods with fewer trees (Kardan et al., 2015). In addition, they discovered lower rates of cancer, diabetes, heart conditions, and mental health illnesses among residents of well-treed neighbourhoods.

Sandifer et al. (2015) reviewed previous research linking human well-being with nature and concluded that the importance of biodiversity to human well-being was immense. They looked at specific physical and mental health outcomes and found that residents living in neighbourhoods with ample natural areas lived longer and experienced lower rates of "anxiety and depression (especially), upper respiratory tract infections, asthma, chronic obstructive pulmonary disorder (COPD), severe intestinal complaints, and infectious disease of the intestine" (p. 6) when compared with individuals lacking in nature contact.

As children's lives are largely dominated by screens designed to demand their attention, they have little opportunity to be outdoors and develop a sense of autonomy; as a result, they may feel they have less control over their lives. Without such opportunities, the future generation of little scientists may develop a disconnected relationship with nature, affecting both their capacity to wonder at the mysteries of the natural world and the many benefits of a life integrated in nature.

Broadly speaking, the teaching of science may be divided into three categories:

- learning of knowledge disseminated by scientists
- learning the ways of scientific thinking
- learning the practices of scientific inquiry.

Many students of science have experienced an emphasis on the learning of knowledge—often in the form of reading a textbook and reproducing the knowledge disseminated in it—with little to no emphasis on ways to think scientifically and on the practices of scientific inquiry. However, science curriculum documents across Canada strongly promote a focus on the development of practices of scientific inquiry. For example, according to Manitoba's science curriculum, a central goal of science education is to "encourage students at all grades to develop a critical sense of wonder and curiosity" (Manitoba Education and Training, 1999, p. 1.2). This chapter focuses on an approach to science education that provides students opportunities for scientific inquiry in a way that fosters the development of capabilities required to live a flourishing life. The following section will address how a particular pedagogical approach to science education may address a disconnected relationship with nature and contribute to human well-being.

10.3 What Is a Capabilities-Development-with-Nature Pedagogy?

I developed the Capabilities-Development-with-Nature (CDWN) approach to teach science because of the possibilities for impacting the development of student capabilities that contribute to their well-being (Link, 2018). The CDWN approach is an integration of the Reggio Emilia Approach® (for complete information on this approach and its inception, see https://www.reggiochildren.it/en/reggio-emilia-app roach/) and outdoor/science education. As a participatory approach, the needs and capabilities are determined by the stakeholders (i.e., elders, teachers, principals, parents, and students) through focus group meetings and interviews, then the teacher designs and creates opportunities for students to develop and enact these needs-linked capabilities. The CDWN approach recognizes that nature can provide a rich source of experiences for the development and enactment of needs-linked capabilities. The capabilities identified in the author's study (Link, 2018) include:

- to listen and respond to others' questions and ideas about nature
- to ask questions about the natural world

- to explore one's own questions and ideas about nature
- to make choices about what to create
- to appreciate and care for nature
- to connect to nature

The Reggio Emilia Approach respects children's imagination and intelligence; it sees the classroom environment as a *third teacher*. The Reggio Emilia programs originated in summer camp settings in Italy (Moss, 2016), and many Reggio Emilia-inspired approaches carry on the practice of integrating the outdoors into project work (e.g., Fraser, 2012). Following this tradition, in the CDWN approach the third teacher is the local natural environment. Teachers orchestrate *provocations* (i.e., points of interest to spark curiosity) during outdoor adventures with their students, who are then provided with opportunities to develop the capability to ask questions and voice ideas about what they encounter. These questions and ideas are recorded by the teacher and may form the basis for a student-led investigation.

The development of the capability to ask questions about the natural world may lead to or overlap with other capabilities. As students are provided opportunities to ask questions or voice ideas, they are afforded opportunities to develop the capability to explore, discuss, and listen to others' ideas and questions they have about nature. In the CDWN approach, children are offered opportunities to develop the capability to care, appreciate, and connect with nature through stewardship or experiential activities, such as nature meditation, that build a bond between child and nature. Finally, teachers provide opportunities for students to develop creative capability.

10.4   Theoretical Influences

The CDWN approach draws on Max-Neef's (1991) Fundamental Human Needs model, Nussbaum's (2013) Capabilities Approach, and Falkenberg's (2019) Well-Being and Well-Becoming Framework. The following subsections provide a brief outline of these theories and their significance to the study described in this chapter.

10.4.1   Fundamental Human Needs

Needs approaches view human well-being as the satisfaction of needs. Max-Neef (1991) offered a list of nine fundamental human needs: subsistence, protection, affection, understanding, participation, idleness, creation, identity, and freedom. These needs are addressed by needs satisfiers. The fulfillment of any of the needs contributes to well-being, while a deficiency in any of the needs results in a poverty of that aspect of well-being. The interconnectedness of the needs is such that a satisfaction of one need may potentially lead to the impoverishment of another. In conjunction with the capabilities approach discussed below, Max-Neef's Fundamental Human Needs model provides a foundation for the study of the impact of a pedagogical approach on student well-being.

10.4.2   Capabilities Approach

Nussbaum's (2013) ten capabilities—"life; bodily health; bodily integrity; senses, imagination, and thought; emotions; practical reason; affiliation; other species; play; and control over one's environment" (pp. 23–24)—were presented as a proposal that should be revised and rethought over time. Nevertheless, Nussbaum contended that this collection of capabilities points to what each individual needs in order to meet the minimum starting point for a good life. Nussbaum (2013) offered the Capabilities Approach through the lens of social justice and equitable opportunities for human growth and development. What opportunities or freedoms are equitably available to individuals in order that they may develop and enact agentic capabilities? Nussbaum emphasized the necessarily agentic nature of the capability, that is, the individual has freedom and control over the capability. While the opportunity to develop capabilities toward well-being is made available, people can decide for themselves whether or not they want to enact that capability in a given situation. This emphasis of choice and agentic control is an important feature of this approach. Meeting basic human needs that contribute to well-being, and allowing for agency over the enactment of the capabilities relevant to a need, is a matter of social justice or an equitable distribution of opportunities necessary to live a good life. For example, a child may have a natural ear for music (an individual characteristic) but no external opportunity to experience or learn a musical instrument due to conditions of poverty (external features). For certain agentic capabilities to develop, opportunities must be available.

Schools should be seen as places where opportunities abound for children to grow and develop. The nature of the school will dictate how narrow or broad and how explicit or hidden their opportunities to develop capabilities will be. For instance, the capability of caring for and appreciating nature contributes to the fulfillment of the human need to give affection, in this case, with regard to nature. Teachers provide students opportunities in the outdoors to practice this capability—for example, students suggest the removal of litter from a pond in order to preserve a livable habitat for ducks and other animals—including the act of planning and reflecting on this action and the freedom to participate or refrain.

10.4.3   Well-Being and Well-Becoming Framework

Building on Max-Neef's Fundamental Human Needs model and Nussbaum's Capabilities Approach, Falkenberg (2019) offered a framework for the conceptualization of human well-being called the Well-Being and Well-Becoming Framework (WB2-Framework). This framework is characterized by a systems perspective that includes people as "bio-psychic systems … [and] as social actors of social and ecological systems" (p. 4). This conceptualization necessarily "requires the consideration of the social-cultural context in which an understanding of well-being and well-becoming is needed" (p. 5). The WB2-Framework is an integrated approach to understanding well-being and well-becoming, drawing from many disciplines to arrive at a holistic understanding of what it means to live well. A key feature of this framework that the CDWN approach adopts is the notion that human needs must be met in order for humans to live a flourishing life (i.e., a life of well-being).

Another core characteristic of the WB2-Framework that is particularly relevant to the idea of an education for well-being is the notion that humans are constantly "becoming" (Falkenberg, 2019, p. 13). Such becoming is the result of the ongoing interaction between a human being and its environment. The dynamic aspect of the interaction is "integral to understanding the quality of a person's present state (well-being)" (p. 5). Falkenberg said that "*well-becoming* expresses the dynamic aspect of well-being and *well-being* expresses the momentary state of well-becoming" (p. 14). Nussbaum's Capabilities Approach and Max-Neef's Fundamental Human Needs approach to human well-being, brought together by Falkenberg's WB2-Framework, provides a foundation for the CDWN approach to science education.

10.5   Methods

My research addressed the central question: How does a particular approach to outdoor learning impact student wellbeing? The study progressed through three phases: identification of capabilities, data collection through interviews and observations, and evaluation of data that reflected opportunities, in an outdoor learning context, to develop capabilities that may contribute to student well-being.

In the preliminary phase of this study, a focus group with 16 stakeholders (eight teachers, five parents, two students, and the principal) identified the needs and needs-linked capabilities that served as the foundation for the design of the CDWN approach. The following two questions guided the identification:

- What needs related to student well-being are you interested in investigating?
- What student capabilities are required to fulfil those needs?

The identified capabilities were assessed through the analysis of observation field notes and interview transcripts, which were the data sources for the remainder of the study.

10.5.1 Setting and Participants

An elementary school in the Canadian prairies was selected for study; it was considered an exemplary model of an effective outdoor learning approach. Of the eight classroom teachers who agreed to participate, Ms. Carson, Ms. Berry, and Ms. Jensen consented to subsequent classroom observations and interviews. The remaining five teachers consented to interviews. Ms. Carson, a Grade 1/2 teacher, was a central focus in this research and I conducted the majority of my nine outdoor observations with her class. Teacher colleagues and parents, as well as the principal, referred to Ms. Carson as a model example of the Reggio Emilia-inspired outdoor education approach and recommended that I focus on her teaching.

Once the three classrooms were established, the parents of the children in these classrooms were emailed by the principal to obtain consent for observations and interviews with their children. In Ms. Carson's classroom, 57% of parents consented to their child's participation, in Ms. Berry's classroom, 56% consented, and in Ms. Jensen's classroom, 28% consented. Once parents had provided consent, I arranged observation times that were convenient for the teachers. Although all students were present for classroom observations, in my field notes I recorded only the interactions and behaviours of those students whose parents had given consent. Children whose parents had given consent were asked if they agreed to participate in interviews.

10.5.2 Procedures

I used observations and interviews as the main data sources. Throughout nine observations conducted in April, May, and June 2017, I looked for examples of opportunities for students to develop and enact the capabilities identified by the focus group in January 2017. I conducted two school-based observations for context, four observations in a Grade 1/2 split class, and three observations in a Kindergarten class; two of those observations included a combined outdoor excursion with other classes.

I interviewed eight teachers (three group interviews), five parents (three separate group interviews), two children (with their parents present), and the principal using a distinct interview protocol for each stakeholder group. For example, in the interview protocol for teachers, under the subheading, *Understanding*, I asked:

- What is an important capability that you think is required to fulfill the need for understanding?
- How might [the CDWN approach] support the development of the capability?
- What have you seen to back up this claim? What activities have the students been involved in within [the CDWN approach] that would support the development of the capability you mentioned? What opportunities does this program afford in the development of this capability? What student responses have you witnessed that indicates that this capability is being developed? (Link, 2018, p. 258)

By listening for key ideas, words or evolving themes, I used each successive interview to build a deeper understanding of a common approach taken to outdoor learning and looked for evidence of the opportunities provided to students to develop capabilities that support well-being.

Member checking of interview transcripts and determining reliability through patterns aided in the triangulation of data and helped offset observer bias (Merriam & Tisdell, 2016). The data from observations and interviews were categorized, coded, and assessed for importance. Interview transcripts and observation notes were organized into data units with reference to the identified capabilities. I then searched the data units for examples of opportunities to develop and enact the capability and assessed those opportunities according to four criteria.

Criterion 1—Frequency: Are there many opportunities to develop and enact the capability?

Criterion 2—Integration: Is the capability valued in such a way that it is developed and enacted on a consistent and frequent basis?

Criterion 3—Participation: Are all or most of the students involved in the development and enactment of the capability?

Criterion 4—Purpose: Are the opportunities to develop and enact the capability meaningful and substantial? Do the students have agency over how, when, or how long to enact the capability?

10.6   Results

Based on the stakeholders' selection of specific needs and capabilities, I analyzed the interview and observational data for the provision of opportunities for students to develop and enact the identified capabilities. I also evaluated the positive impact of the CDWN approach to develop and enact these capabilities and ultimately the well-being of the students. The remainder of this chapter will address the six capabilities identified by the participating stakeholders (Table 10.1). Each capability in the CDWN approach is described, examples of each capability from the findings (i.e., classroom, outdoor observations, and semi-structured interviews) are presented, and key characteristics are discussed.

10.6.1   Participation: To Listen and Respond to Others' Questions and Ideas

To meet the need of participation following the CDWN approach, the teacher provides opportunities in nature for students by developing the capability to share questions and ideas and to listen and respond to others' questions and ideas. An important feature is the teacher's beliefs about the value of students' voices. Providing space

Table 10.1   Stakeholder-identified needs, capabilities, and examples

Need	Capability	Example
Participation	To listen and respond to others' questions and ideas	Listening and responding to ideas and questions during a cankerworm debate (should the cankerworms here be destroyed to prevent them from destroying this tree?)
Understanding	To ask questions about the natural world	Observing and asking questions about forest tent caterpillars on the window ledge of the school
Freedom	To explore one's own questions and ideas about nature	Exploring questions about water insects that were inspired by a trip to the local pond
Creation	To make choices about what to create	Representing a garden gnome village in the forest built with elements of nature
Affection	To appreciate and care for nature	Appreciating bees and caring for spiders
Identity	To connect to nature	The peaceful heart meditation practice at the forest, pond, and other local sites

for students to voice their opinions, ideas, perceptions, stories—as well as ensuring opportunities for classmates to listen and respond to those articulations—is fundamental to the development of the capability. For some teachers, it may challenge the very core of what they think it means to be a teacher, that is, to talk while students listen.

10.6.1.1   Examples and Key Characteristics

One example of developing the capability to listen and respond to others' questions and ideas involved an unplanned student-directed and teacher-facilitated cankerworm debate. At the end of an outdoor activity concerning the Seven Sacred Teachings of the Anishinaabe, a few students noticed some cankerworms ascending a young tree and began to discuss plans to exterminate them. This caught the attention of a few other students who expressed their view that all life was important and that they should not kill the cankerworms. Their teacher, Ms. Carson, recognized this moment as an opportunity to develop the listening/responding capability and facilitated a short discussion, helping them articulate their viewpoints. In the end, the discussion was temporarily paused and continued indoors later in the afternoon. Other examples included opportunities to discuss and plan student-led initiatives, specifically, sharing a nature walk/litter clean-up, a school/classroom garbage measurement, an aquarium/animal habitat for the classroom, a water stewardship action and litter clean-up planning, and ideas initiating and regarding a rain inquiry.

10.6.2   Understanding: To Ask Questions About the Natural World

In the Reggio Emilia Approach, there are three teachers: adults, children, and the physical environment (Edwards et al., 2012). The learning environment is thoughtfully designed to provoke curiosity and imagination and provide opportunities for children to question, explore, play, and create. A typical Reggio Emilia learning environment includes well-organized areas for individual and quiet engagement, small- and large-group discussion, inquiry and creative work, and imaginative play (Edwards et al., 2012). Similarly, in the CDWN approach, the learning environment (i.e., the third teacher) extends to the natural environment. Children's questions and interests can be recognized and reflected in the way in which engagement in the outdoors is designed. Teachers document questions the students have about the world. These questions may form the basis of inquiry projects and such student-initiated inquiries are often more meaningful for students. By extension, the experiences in the outdoor environment can act as provocations for inquiry and creative work and play in the indoor environment. What makes the CDWN a science education approach is the intentional engagement with nature to provoke curiosity.

10.6.2.1   Examples and Key Characteristics

Ms. Carson described being outdoors as an opportunity to accelerate learning and provide opportunities to ignite curiosity. In nature, it is possible to encounter unique learning experiences whether planned, hoped for, or unexpected. For example, one morning as the class was finishing their garden plan development, some students discovered a cluster of forest tent caterpillars on an outside window ledge of the school. Ms. Carson produced the pad of paper and pen that she always carried around in her back pocket while listening to their conversations. At first, the students were content just to observe and talk about the tent caterpillars. After a time, they were asked to return to gather their garden plans. One student approached Ms. Carson with a question about the life cycle of tent caterpillars; she replied that this was a fabulous question and wrote it down on her pad of paper. This question was revisited later in the classroom and formed the basis for a student-led inquiry.

10.6.3   Freedom: To Explore One's Own Questions and Ideas About Nature

A drive to explore questions and ideas about nature may be motivated by the wonder and curiosity that children have in their encounters with the richness of nature. Through intentional engagement with nature and the freedom to explore their own questions and ideas, children may experience a sense of empowerment.

Being provided with opportunities to explore their questions indicates to students that their questions are important—that recognition is empowering. The CDWN approach makes the development of critical thinking, innovative, and literacy skills meaningful because these skills are developed and enacted in an authentic setting: the natural world. Writing, thinking, and creating something drawn from student-generated questions and explorations about nature can be meaningful and exciting work for them.

The way in which the CDWN approach provides opportunities to explore questions and ideas includes cultivating an environment in which students trust teachers to support them in their endeavours, struggles, mistakes, as well as in the excitement of their explorations. This type of environment helps children feel cared for, supported, and valued. While this supportive attitude may be present within the walls of a classroom, it is more authentic if questions about nature are explored first-hand in nature. In order to explore questions and ideas about a farm, for instance, going to a farm is arguably more engaging than reading about it in a classroom.

10.6.3.1 Examples and Key Characteristics

The impact of the CDWN approach to providing opportunities to explore questions and ideas was described as positive with regard to children's confidence in themselves and their comfort level with making a mistake. Ms. Carson was alive to the types of opportunities offered in nature. Life outside of the classroom is certainly more unpredictable. While Ms. Carson always ventured out-of-doors with a plan—anticipating, preparing for, and mitigating any possible risks, she kept a watchful and open mind with regard to the students' responses and queries in what they encountered. This approach allows for opportunities to explore questions and ideas that students have about nature. I also observed this openness in other classes as well as during interviews with parents and teachers in their descriptions of teachers' openness to educative moments.

One example of the opportunity to explore one's own questions and ideas about nature involved an outdoor study of a local pond. This experience included exploring questions about pond insect identity through the use of field guidebooks and resource insect identification cards. The central purpose of such nature studies is to provide opportunities for students to ask and then later explore questions and ideas about what they encountered. Ms. Carson took note of their questions as potential sources of inquiry for future study.

10.6.4 Creation: To Make Choices About What to Create

The teacher provides opportunities in nature for students to meet the need of creative work by developing the capability to make choices about what to create. The Reggio Emilia concept of "the hundred languages of children" (Edwards et al., 2012, p. 1) is directly linked to the creative choice capability. The hundred languages of children suggests that children should be encouraged to express themselves through whatever means they can, for example, sculpting, building, painting, drawing, etc. Because children communicate and learn through various ways, a variety of materials should be provided for discovery and expression of their questions and understandings and of what they feel or imagine. As teachers provide these conditions, students may develop and enact the capability to make choices about what to create.

10.6.4.1 Examples and Key Characteristics

An example of developing the capability for creative choice involved inviting the students to create various representations in a forested area using elements from nature, such as sticks, rocks, and clay. These representations included fairy tale worlds, gnome villages, and other small magical worlds. In the outdoors, the students arguably had more choices about what kind of world to create and more choices of how to create it by using varied, non-uniform materials. Being in the forest seemed to feed their imagination. For instance, when they built a gnome village, many of them, through their sense of fantasy and imagination, pretended that there were in fact gnomes waiting to move into the newly constructed village. A second example came in the design of garden blueprints by the students which led to the planting of outdoor gardens. The creation of the blueprints for the garden, as with the construction of the gnome village, required meaningful choices. Upon completion of the garden and magical worlds projects, the students saw an end product where they made significant decisions in the planning and execution.

Numerous activities observed and described by teacher, parent, and student participants regarding creative choice share the characteristic of using the outdoors to spark the students' imagination. Many of the their art and writing projects, whether created in the classroom or outside, were inspired by their outdoor adventures, including hikes in the forest, explorations in the pond, or encounters with forest tent caterpillars. These initiatives were instigated by the students; the activities linked to these initiatives were student-directed and represent an example of engineering design.

10.6.5 Affection: To Appreciate and Care for Nature

Appreciating and caring for nature—arguably the most important capability for a child to develop from the perspective of living sustainably—may take many forms. Students may develop and enact an appreciation and care for nature through their teacher's guidance and modelling, helping them empathize with their classmates and with animals encountered both outdoors and in the classroom. For a young child, to care for nature may include recognizing that a pond is home to many plants and animals; consequently, it is important to tread carefully when exploring the shores of a pond. To appreciate nature comes quite naturally for many children and can be as simple as sitting in awe of a woodpecker hammering away on a tree or as sophisticated as recognizing the interconnectedness of life systems. Ms. Carson talked about the importance of being flexible and responsive to educative moments, especially moments that lend themselves to feelings of appreciation and wonder in the natural world and the opportunities that these moments bring for asking and exploring questions. A sense of wonder appears to be a significant factor in the desire to explore questions and ideas about the world. Again, this sense of wonder and curiosity, and the opportunity for the questions that they afford, is more easily available when children are taken into nature.

10.6.5.1 Examples and Key Characteristics

Opportunities to develop and enact the capability to appreciate and care for nature were provided through a gratitude practice. Before leaving a natural site, such as a pond or forest, the students took a moment to offer gratitude to the land through a silent reflection. Awareness activities—sitting and simply listening in silence to the sounds of the birds, leaves, wind, etc. upon arrival at their destination in the forest—provided opportunities to develop the capability to appreciate nature.

A wonder wagon, brought along on outdoor adventures, provided opportunities to develop the capability for appreciating nature. The wonder wagon consisted of many tools for examining life and for capturing and appreciating the beauty that was encountered: magnifying glasses and clear containers; field guides to identify insects, mammals, birds, and plant life; artistic materials such as paper, pencils, crayons, paints, paintbrushes, clay, and Plasticine™. At the beginning of the year, teachers spent time facilitating discussions around caring and respectful engagement in the outdoors. Crucially, ideas such as *Don't harm the animals or their habitat* came from the students through teacher facilitation. Care and appreciation for nature will sometimes lead to student-initiated action, such as litter patrols. These actions, whatever they may be, were again student-generated and supported by teachers through facilitated discussion, help with planning, and dealing with logistical concerns.

10.6.6 Identity: To Connect to Nature

In this final capability of the CDWN approach, the teacher provides opportunities in nature for students to meet the need of identity by developing the capability to connect to nature. This capability may lead to an *I am with nature* identity. The feeling of being connected to other life and natural elements undoubtedly includes human connection but may also extend further to include animals, rocks, plants, entire ecosystems, and so on. Experiencing a connection to nature by being in nature itself helps students reflect on their identity in relation to the natural world. Such an identity of connectedness is supported by providing them with opportunities to feel connected to one another in nature as well as in the classroom.

10.6.6.1 Examples and Key Characteristics

As students developed and enacted the capability to connect with nature, they identified themselves as stewards for nature. This identity led them to take action to keep clear of litter the natural habitat of animals in the local lake, forests, ponds, and their backyard. In this type of connected environment, children support one another; as in the story described in an interview by one of the teachers of a girl who shared her feelings about her parents' divorce and the immediate response of the students to comfort her as she cried. As in all ecosystems, elements of life exist in relation to one another and cannot exist in isolation; so too in the classroom, students may flourish in their connection to one another and to the wider community of life and life-giving elements.

Students were provided opportunities to artistically represent the connections they felt to nature when they were outdoors during local adventures and field trips further abroad. This capability to connect with nature might be developed through activities such as the gratitude practice and awareness activities. The connection may also be developed through student-initiated actions, for example, planting and sustaining a tree, raising money for water stewardship, or thoughtful harvesting of plants (i.e., picking dandelions but leaving other species).

10.7 Discussion

Teachers of the CDWN approach can provide opportunities in a natural environment to develop students' capabilities that are linked to well-being. Many of the experiences designed by the teacher showcased in this chapter offer a multitude of opportunities to engage and connect with the natural world. Further, the capabilities identified in the CDWN approach provide experiences that are linked to science skills. The capability to listen and respond to others' questions and ideas about nature is connected to science communication goals, a foundational aspect of scientific inquiry. The capability to make choices about what to create is linked to engineering and design. Finally, the capabilities to appreciate and care for nature and to experience a connection to nature are linked to science values and attitudes.

There are some overlaps among the capabilities in the CDWN approach. The approach's provision of opportunities for student development was described on numerous occasions by teachers, parents, and the principal during interviews and the focus group meeting as well as witnessed repeatedly during the observation sessions. The freedom to explore one's own ideas and questions about nature seemed to overlap with the other capabilities to a substantial degree perhaps due to the notions of childhood that are fundamental to the CDWN. A central notion of the CDWN approach referred to here is that of agency, that is, an individual's ability to act on one's will. In the Reggio Emilia philosophy, children should be provided with the agency to, for example, explore their ideas and questions about the world.

The rationale for providing opportunities to explore students' ideas and questions stems from the teachers' belief that children are naturally inquisitive beings who require opportunities to explore the questions and ideas they have about nature. In developing a connection with nature, teachers also provide opportunities to explore action-orientated ideas (e.g., a student-driven litter-less lunch initiative). The freedom to go outside and explore questions and ideas about nature was described by Ms. Carson as student-led and authentic, not dictated by schedules or routines. As a result, these students had a substantial degree of agency in satisfying their curiosity. Experiencing a degree of choice and control can positively impact one's sense of well-being.

As an example that encompasses all six capabilities, consider a teacher providing the provocation of a disposable cup designed for a hot beverage, such as hot chocolate. The students may voice questions and ideas about how the heat from the hot chocolate dissipates then listen and respond to each other's questions and ideas. With the teacher's support, students may ask questions about heat loss that could form the basis for an investigation. Students may then be encouraged, based on their investigation into heat loss, to create an insulated cup to slow the rate of heat loss. The values-based capabilities to appreciate and care for nature and to experience a connection to nature may inform the choice of materials that are used to construct the cup. Children enacting the capabilities to appreciate and feel connected to nature might select sustainable materials (perhaps an edible cup) or challenge the notion of disposability itself.

Ms. Carson described the importance of modeling an appreciation and wonder for the big and little moments that can be observed on a nature walk. She was keenly aware that she cannot compete with the brilliance of a bald eagle soaring overhead, even if she was on the cusp of making an important point. In these moments, she patiently stops and then models and reflects the excitement and wonder that the students experience. This modeling was especially important during the first few weeks of the school year, because this time is the most challenging, especially for students who are new to the school. For example, Ms. Carson described some students who had not had as much opportunity to appreciate or care for nature demonstrating revulsion to beetles and expressing a desire to squash the insects. By modeling an appreciation and care for life, she prepared the children for opportunities to develop and enact the capability to value and act as stewards within the natural world.

10.7.1 Key Characteristics of the CDWN Approach

One of the key characteristics of the CDWN approach is the recognition of the natural world as a place of wonder and mystery to spark imagination, curiosity, thought, and action. Whether nature is recognized as a means to excite children to engage in creative work—as a spark for a discussion, a question, an exploration, an ethical action—or as an opportunity to reflect on identity as a part of nature, the natural world is a central part of this science education approach. For teachers, the natural world can be a source of provocations that may lead to students' inquiries, artistic endeavours, projects, or perhaps even a deeper awareness of their space within the web of life.

Another key characteristic involves looking for and recognizing opportunities in nature to develop and enact the corresponding capabilities. Special moments in nature can provide an opportunity for students to develop a range of capabilities. Sometimes, as in the case of spotting a soaring bald eagle on a walk to the pond, the opportunity means modelling an excitement and wonder at the sight of this magnificent being. Other situations might call for recognizing and recording questions that students have about, for example, a ladybug that they have discovered. The ladybug becomes the provocation that may, in turn, lead to a class-wide inquiry project into ladybugs or some aspect of ladybugs about which they are curious.

Another important characteristic is the teachers' attunement to children as inquisitive beings. With this perspective, children's questions are heard, valued, and explored. The importance placed upon respecting the child's inquisitiveness and need for expression informs how educational opportunities are enacted. Children's curiosity, creativity, and sense of self may thrive in nature.

Although being immersed in nature is a key characteristic, indoor work can also support the development and enactment of capabilities. What is experienced on an outdoor adventure may be the source of inspiration for the next day's indoor lesson. For example, an indoor writing activity could be a reflection on the previous day's experience of lying down in the forest and staring up at the treetops.

The final key characteristic of CDWN is agentic control, which permeates all of the capabilities. Teachers provide opportunities for their students to make decisions including the planning and execution of projects and investigations. Providing children with the agentic control over what they do and how they do it, within an open and flexible framework, is a powerfully engaging feature of this science education approach that may instill a sense of empowerment and confidence.

10.8   Limitations and Future Directions

While I would suggest that readers consider the adoption of appropriate aspects of CDWN approach into their contexts, the findings of this case study evaluation may only be strictly applied to the studied school (Link, 2018). Further, the scope of the study did not include a consideration of the contributions that family and community surely make towards child well-being. The evaluation was limited to unstructured observations in three classrooms and to qualitative interview data. I also acknowledge my own bias as a proponent for nature-based education and as a former teacher who had made modest attempts at integrating nature-based education into my own teaching practice. Finally, it should be noted that the lack of student involvement in the development of needs and capabilities represents a shortcoming of this study. There were simply not enough interested children with parents who were willing to allow their children to be interviewed. This shortcoming does contradict the spirit of the Reggio Emilia Approach, which values the voice of children.

The potential for further research in this exciting field is vast. There is a gap that exists in both the literature and in the explicit mandate of schools to support child well-being through the development of appropriate capabilities with nature, although there are schools that are already quietly supporting children's well-being in this way. Further research could identify and study schools that utilize the outdoors to develop capabilities that contribute to child well-being using approaches other than the CDWN. There is also much to be learned by studying schools that provide opportunities for children to develop capabilities that contribute to well-being without prominently utilizing the outdoors (if at all). Research might focus on teachers who provide opportunities to their students to develop important capabilities who are informed by educational theories other than the Reggio Emilia philosophy. With the CDWN and other capability-enhancing approaches, longitudinal research might assess impacts on student well-being into adulthood. Finally, I suggest that further research into the CDWN approach include a greater emphasis on the child's perspective, an aspect that was only touched upon in this study.

10.9 Final Thoughts

Why do we send our children to school? Is it in pursuit of a *good life*? A high-quality education, it might be hoped, will provide the knowledge and skills needed to acquire a trade or profession that offers a stable income; but are schools more than just places to prepare for a life of work and consumption? With increasing rates of depression and anxiety among children, school mandates need to consider the well-being of children (Birmaher et al., 2007; Findlay, 2017; Tong & McLeod Macey, 2017) as we endeavour to live well on a finite planet.

In this chapter, I have articulated how teachers following the CDWN approach can provide opportunities for their students to develop six capabilities important to fundamental human needs. When teachers provide opportunities in nature for children to develop the capabilities to voice, listen, ask, and explore questions about the natural world in the outdoors, these opportunities can (a) help fulfill the fundamental human need for understanding and (b) provide opportunities for developing an inquisitiveness that may lead to researching and communicating information. When teachers provide opportunities in nature to develop the capability to make choices about what to create, they are honouring the creative and intellectual domain of children; this capability is connected to the fundamental human need for creation. When teachers make allowances for children to develop the capabilities to appreciate and care for nature and to experience a connection to nature, the seeds for a relationship with nature are planted; these capabilities are linked to the fundamental human need for affection and identity and may benefit well-being.

The students featured in this chapter were offered the opportunity to engage in meaningful pursuits that explored authentic questions, projects, and connections of interest. The students' practice of science—purposefully crafted and facilitated by the teacher with children's agency in mind—allowed for them to develop capabilities that support well-being. As ecopsychologists attest (e.g., Gupta et al., 2018), providing students with opportunities to appreciate, care, and connect with nature may help to solidify core aspects of the human identity necessary for well-being.

References

Bee, H., & Boyd, D. (2010). *The developing child* (12th ed.). Allyn & Bacon.

Birmaher, B., Brent, D., & AACAP Work Group on Quality Issues. (2007). Practice parameter for the assessment and treatment of children and adolescents with depressive disorders. *Journal of the American Academy of Child & Adolescent Psychiatry, 46*(11), 1503–1526. https://doi.org/10.1097/chi.0b013e318145ae1c

Cacioppo, J. T., Hawkley, L. C., Kalil, A., Hughes, M. E., Waite, L., & Thisted, R. A. (2008). Happiness and the invisible threads of social connection: The Chicago health, aging, and social relations study. In M. Eid & R. J. Larsen (Eds.), *The science of subjective well-being* (pp. 195–219). Guilford Press.

Carrington, D. (2016, March 25). Three-quarters of UK children spend less time outdoors than prison inmates – survey. *The Guardian.* https://www.theguardian.com/environment/2016/mar/25/three-quarters-of-uk-children-spend-less-time-outdoors-than-prison-inmates-survey

Edwards, C., Forman, G., & Gandini, L. (Eds.). (2012). *The hundred languages of children: The Reggio Emilia experience in transformation* (3rd ed.). Praeger.

Falkenberg, T. (2019). *Framing human well-being and well-becoming: An integrated systems approach.* (Working Paper Series, #2.) http://wellbeinginschools.ca/wp-content/uploads/2019/09/WBIS-Paper-No-2-Falkenberg-2019-2.pdf

Findlay, L. (2017). Depression and suicidal ideation among Canadians aged 15 to 24. *Health Reports, 28*(1), 3–11. https://www150.statcan.gc.ca/n1/en/pub/82-003-x/2017001/article/14697-eng.pdf?st=LjgruBEy

Fraser, S. (2012). *Authentic childhood: Experiencing Reggio Emilia in the classroom* (3rd ed.). Nelson.

Gordon, J., & O'Toole, L. (2015). Learning for well-being: Creativity and inner diversity. *Cambridge Journal of Education, 45*(3), 333–346. https://doi.org/10.1080/0305764X.2014.904275

Gupta, R., LaMarca, N., Rank, S. J., & Flinner, K. (2018). The environment as a pathway to science learning for K–12 learners—A case study of the E-STEM movement. *Ecopsychology, 10*(4), 228–242. https://doi.org/10.1089/eco.2018.0047

Kahn, P. H., & Hasbach, P. H. (Eds.). (2012). *Ecopsychology: Science, totems, and the technological species.* MIT Press.

Kardan, O., Gozdyra, P., Misic, B., Moola, F., Palmer, L., Paus, T., & Berman, M. (2015). Neighborhood greenspace and health in a large urban center. *Scientific Reports, 5,* 11610. https://doi.org/10.1038/srep11610

Kern, M. L., Waters, L. E., Adler, A., & White, M. A. (2015). A multidimensional approach to measuring well-being in students: Application of the PERMA framework. *Journal of Positive Psychology, 10*(3), 262–271. https://doi.org/10.1080/17439760.2014.936962

Kharel, M., Sakamoto, J. L., Carandang, R. R., Ulambayar, S., Shibanuma, A., Yarotskaya, E., Basargina, M., & Masamine, J. (2022). Impact of COVID-19 pandemic lockdown on movment behaviours of children and adolescents: A systematic review. *BMJ Global Health., 7*(1), e007190. https://doi.org/10.1136/bmjgh-2021-007190

Lien, A., Sampasa-Kanyinga, H., Patte, K. A. Leatherdale, S. T., & Chaput, J.-P. (2022). Sociodemographic and mental health characteristics associated with changes in movement behaviours due to the COVID-19 pandemic in adolescents. *Journal of Activity, Sedentary, and Sleep Behaviours, 1*(5). https://doi.org/10.1186/s44167-022-00004-2

Link, M. (2018). Nature, capabilities, and student well-being: An evaluation of an outdoor education approach. Doctoral dissertation, University of Manitoba. http://hdl.handle.net/1993/33289

Louv, R. (2011). *The nature principle: Human restoration and the end of nature-deficit disorder.* Algonquin Books.

Manitoba Education and Training. (1999). *Kindergarten to Grade 4 science: Manitoba curriculum framework of outcomes.* https://www.edu.gov.mb.ca/k12/cur/science/outcomes/k-4/index.html

Mannell, R. C. (2007). Leisure, health and well-being. *World Leisure Journal, 49*(3), 114–128. https://doi.org/10.1080/04419057.2007.9674499

Max-Neef, M. A. (1991). *Human scale development: Conceptions, applications and further reflections.* Apex Press.

Mcknight, P. E., & Kashdan, T. B. (2009). Purpose in life as a system that creates and sustains health and well-being: An integrative, testable theory. *Review of General Psychology, 13*(3), 242–251. https://doi.org/10.1037/a0017152

Merriam, S. B., & Tisdell, E. J. (2016). *Qualitative research: A guide to design and implementation* (4th ed.). Jossey-Bass.

Moss, P. (2016). Loris Malaguzzi and the schools of Reggio Emilia: Provocation and hope for a renewed public education. *Improving Schools, 19*(2), 167–176. https://doi.org/10.1177/136548 0216651521

Nussbaum, M. C. (2013). *Creating capabilities: The human development approach.* Harvard University Press.

Sandifer, P. A., Sutton-Grier, A. E., & Ward, B. P. (2015). Exploring connections among nature, biodiversity, ecosystem services, and human health and well-being: Opportunities to enhance health and biodiversity conservation. *Ecosystem Services, 12*, 1–15. https://doi.org/10.1016/j.eco ser.2014.12.007

Seguin, D., Kuenzel, E., Morton, J. B., & Duerden, E. G. (2021). School's out: Parenting stress and screen time use in school-age children during the COVID-19 pandemic. *Journal of Affective Disorders Reports, 6*, 100217–100217. https://doi.org/10.1016/j.jadr.2021.100217

Seligman, M. E. P. (2011). *Flourish: A visionary new understanding of happiness and well-being.* Free Press.

Seppala, E., Rossomando, T., & Doty, J. R. (2013). Social connection and compassion: Important predictors of health and well-being. *Social Research: An International Quarterly, 80*(2), 411–430. https://www.muse.jhu.edu/article/528212

Strasburger, V. C., Jordan, A. B., & Donnerstein, E. (2010). Health effects of media on children and adolescents. *Pediatrics, 125*(4), 756–767. https://doi.org/10.1542/peds.2009-2563

Tong, G., & McLeod Macey, J. (2017, November 14). *Children and youth mental health survey: Getting help in Ontario.* Ipsos. https://www.ipsos.com/en-ca/news-polls/CMHO-children-and-youth-mental-health-ontario

Vézina-Im, L.-A., Beaulieu, D., Turcotte, S., Roussel-Ouellet, J., Labbé, V., & Bouchard, D. (2022). Association between recreational screen time and sleep quality among adolescents during the third wave of the COVID-19 pandemic in Canada. *International Journal of Environmental Research and Public Health., 19*(15), 9019. https://doi.org/10.3390/ijerph19159019

Michael W. Link BEd (University of Manitoba), PBDE (Simon Fraser University), MEd (Simon Fraser University), PhD (University of Manitoba), is an assistant professor of science education in the Faculty of Education at the University of Winnipeg. He taught for 13 years in diverse and marginalized communities and has been teaching at the post-secondary level since 2008, predominantly in the areas of science education, educational psychology, and education for sustainability. Michael's current projects include an investigation of the impact that the ongoing global pandemic has had upon teachers of sustainability and global citizenship, a partnership with a land-based learning centre exploring how teachers integrate land-based learning into classroom curriculum and local settings, and an exploration of a well-being in schools approach that provides opportunities for reflection and engagement with community and local ecological issues.

Chapter 11
Using Wearable GPS Technology to Explore Children's Authentic Interest in Nature

Jesse Jewell, Todd M. Milford, and Christine D. Tippett

11.1 Introduction

Children have a special relationship with the natural world that is unique from that of adults (Chawla, 2020; Green, 2016, 2017; Hughes et al., 2019; Hyun, 2005; Kahn et al., 2009; Louv, 2005). They learn and explore in the natural world using primarily their senses, whereas adults tend to perceive nature based on previous experiences and knowledge (Boileau, 2011; Green, 2017; Hyun, 2005; Kahn et al., 2020; Sebba, 1991; Wilson, 1995). Green (2016, 2017) contended that, in order for children's sensory exploration in nature to be authentic and to reflect their interests, they require a degree of autonomy (Deci & Ryan, 2000). The Yukon Territory (YT) recently implemented the revised British Columbia Ministry of Education (2022) science curriculum that focuses on authentic and student-centered instructional strategies. In this study I leveraged emerging geospatial technology tools to investigate 7- and 8-year-old children's interest in the boreal forest in YT, specifically, where they chose to go and what they did when given autonomy in nature.

The decreasing amount of direct nature contact for children in North America, which has coincided with a period of continued urbanization and subsequent disconnection from nature, has attracted attention and concern from educators (Louv, 2005;

J. Jewell (✉)
Yukon Wild School, Whitehorse, Yukon Y1A0C2, Canada
e-mail: jesseojewell@gmail.com

T. M. Milford
Faculty of Education, University of Victoria, STN CSC, Victoria, BC V8W 2Y2, Canada
e-mail: tmilford@uvic.ca

C. D. Tippett
University of Ottawa, Ottawa, ON K1N 6N5, Canada
e-mail: ctippett@uottawa.ca

© The Author(s), under exclusive license to Springer Nature Switzerland AG 2023   187
C. D. Tippett and T. M. Milford (eds.), *Exploring Elementary Science Teaching and Learning in Canada*, Contemporary Trends and Issues in Science Education 53,
https://doi.org/10.1007/978-3-031-23936-6_11

Rosenow & Bailie, 2014). The forest school movement is an example of how educators are reacting to this disconnect. Although there are barriers to accessing off-site learning environments, research has revealed that teachers' perceived value of off-site learning is not one of those barriers (Ernst, 2014; Ernst & Tornabene, 2012; Rickinson et al., 2004). Instead, barriers for primary education centre on issues of access, such as walking distance, time, liability, and weather (Ernst, 2014; Rickinson et al., 2004).

All of YT's 23 elementary schools are within walking distance to the northern boreal forest, making this environment a natural extension of the physical school where students play and learn. YT elementary schools have been developing outdoor learning spaces in recent years, for example, a fire pit with a circle of benches or a canvas wall tent with a woodstove. Outdoor settings can be accessed by primary students in all seasons, with an emphasis on unstructured play in a natural setting (Ernst, 2014; Ernst & Tornabene, 2012). There is room to explore teaching and learning in this environment. In my current position as the Experiential Education Curriculum Consultant at Yukon Education, I have a vested interest in learning more about student interest in nature. Questions that emerge include:

- What do young learners wonder about in the boreal forest?
- How do they approach learning in this environment?
- How can teachers facilitate agency-driven learning?

In the boreal forest, the curriculum is written in the landscape. The biotic and abiotic features of the landscape constitute the content. This curriculum, in combination with students' natural affinity toward play-based exploration of the outdoors, has potential for their engagement and motivation.

11.2   Literature Review

This literature review focuses on two areas of research concerned with children's autonomy in nature. The first area explores students' voices in relation to their movement in nature. The second area concerns the use of wearable technology for data collection while exploring nature.

11.2.1   Student Voice

Honouring student voice is a growing trend in research with young children (Caiman & Lundegard, 2014; Green, 2016, 2017). Young children's learning outdoors occurs in a spontaneous way that necessitates giving them space, time, and autonomy in these environments (Prince et al., 2013). Waters and Maynard (2010) investigated what specific elements in the outdoor environment draw 4- to 7-year-old children's attention when given a chance to explore in the same park on several occasions; they concluded that it was the trees, rocks, and hills and recommended that unkept and natural spaces be made available for exploration. The richness of such environments should not be underestimated.

Kalvaitis and Monhardt (2012) investigated 6- to 11-year-old students' interest in nature by analyzing the drawings they created following their experiences. This study included 10 classrooms from the Rocky Mountain region of the United States. Students were asked to draw a picture of themselves doing something in nature that they valued and then write about the picture to explain their relationship with nature. This methodological approach aimed to capture their interactions *in* nature rather than *about* nature. Students in the lower grades shared more about nearby nature (e.g., watching bugs, picking flowers) whereas older students depicted themselves in more outdoor events (e.g., hiking) or more complex processes (e.g., thinking or feeling). Kalvaitis and Monhardt recommended that curriculum designers and educators pay more attention to specific developmental aspects of student–nature relationships in order to benefit from student interest.

Cheng and Monroe (2012) focused on Grade 4 students' connections to nature and interest participating in nature-based activities. Their survey study was conducted in Florida with 1,432 of 5,550 responding. Findings revealed that students' connections to nature were significantly correlated with nature near their home. Additionally, their connections to nature were correlated with other variables such as previous experiences in nature, knowledge of the environment, and family values.

Ghafouri (2014) observed Ontario Kindergarten students' engagement with the natural environment, focusing on how engagement with nature contributed to learning experiences and how free-choice affected learning. Ghafouri contrasted two distinctly different outdoor experiences: an open-ended exploration of nature close to the students' school and a trip to a local farm. The open-ended exploration inspired many questions from the students, who were also co-constructors of knowledge and directors of their learning during and after the experience. In contrast, although the students were reported to enjoy the farm visit, there was transfer rather than co-construction of knowledge. Ghafouri concluded that attention should be paid to student autonomy when advocating for nature-based experiences with students.

Green (2013) conducted a study of 12 Kindergarten students' connection to place in the Rocky Mountain region of the United States and reached a similar conclusion to Ghafouri regarding student autonomy. Data were collected from student-led tours of special places inside their homes, representational models of these places, guided classroom conversations, and informal conversations. Green's findings revealed that students prefer to have several special places rather than just one. Different special places often had different uses such as exploring or hiding. Students were attracted to environments that provided a sense of autonomy and control through claiming and constructing, creating rules, being creative and imaginative, and exercising environmental competency. Green concluded that we can gain valuable insight into children's interests by listening to them talk about their special places.

The key findings of these studies about students' voices revealed some trends. Across several studies, children displayed an affiliation with nature that was related to sensory exploration and centered on play, including imaginative and landscape-oriented play; for example, trees afforded climbing, dense vegetation afforded hiding games (Green, 2016). Other results revealed children's fascination with both living and non-living aspects of nature, for example, animals and loose parts such as sticks and rocks (Green, 2013; Kalvaitis & Monhardt, 2012; Waters & Maynard, 2010). Additionally, children's autonomous engagement with the outdoors promoted collaboration among peers (Ghafouri, 2014).

11.2.2   Technology and Nature-Based Research

The literature reviewed in this section highlights studies involving young students in nature where data were collected using wearable technology including cameras and tracking technology such as GPS devices. Green (2016) conducted a study with 31 3- to 6-year-olds enrolled in a forest-based summer program in Alaska where data were collected using wearable cameras. Students were given autonomy to explore and experience free play in the forest while the cameras captured where they went, what they did, and what they said. Analysis of the data revealed that these students' interactions with nature were imaginative and socially constructed with their peers. Students paid close attention to aspects of ecology such as insects and participated in gross motor activities such as tree climbing. This process offered an insider's perspective into children's experiences in nature; they were able to articulate what the camera footage meant to them rather than the videos being interpreted solely by the researcher.

Loebach and Gilliland (2010) explored the perceptions of 16 children between the ages of 7 and 9 years in London, Ontario. Children were equipped with maps and digital cameras to record neighbourhood features. GPS technology was utilized to track the routes taken by the children that were then mapped, permitting spatial relationships to be analyzed. The authors noted that the accuracy of GPS data may have been affected by cloud cover, tree canopies, and/or proximity to buildings.

Fjørtoft et al. (2009) organized a study in southern Norway using wearable GPS units and heart rate monitors to track 6- to 7-year-old students' movement patterns in two schoolyards. The two school landscapes afforded different opportunities for students; one was comprised primarily of a small soccer field and asphalt-covered areas with play structures, the other included a forested area. Despite the differences in landscape, physical activity levels were similar for both schoolyards. The wearable technologies enabled the researchers to reduce the impact of an adult's presence on student behavior and autonomy, which strengthened the authenticity of the data collected. However, the GPS data were not always accurate enough for the researchers to differentiate among pieces of equipment in close proximity.

The affordances of wearable technology includes its durability (Green, 2016), the reduction of disruptions created by adult presence, and the ability to obtain a more realistic glimpse of student interest in nature (Fjørtoft et al., 2009; Green, 2016). However, there were some disadvantages noted by researchers, particularly issues of accuracy (Fjørtoft et al., 2009; Loebach & Gilliland, 2010).

11.2.3 Summary

The key findings in the reviewed studies on student voice, autonomy, and the use of wearable technology revealed some consistent trends. For example, children displayed a similar affiliation with nature that was led by sensory exploration and revolved around (a) play, including imaginative play and play that the landscape afforded (e.g., trees afforded climbing, dense vegetation afforded hiding games), and (b) a fascination with living and non-living aspects of nature, particularly animals or evidence of animals and loose parts (e.g., berries, sticks, rocks; Green, 2013, 2016, 2017; Kalvaitis & Monhardt, 2012; Waters & Maynard, 2010). Children's engagement with the outdoors promoted collaboration among peers that emerged when they were given autonomy (Caiman & Lundegard, 2014; Ghafouri, 2014; Green, 2016, 2017). The affordances of wearable technology reduced the disruption created by adults and allowed an authentic glimpse of student action in outdoor environments (Caiman & Lundegard, 2014; Fjørtoft et al., 2009; Green, 2016, 2017; Loebach & Gilliland, 2010). There is still room to expand the body of knowledge on how educators can assist primary students in discovering or rediscovering their interest in nature and, more specifically, on how best to understand student interest in the boreal forest in the YT using wearable technologies. My research question was: Where do students go and what do they do when given autonomy in nature?

11.3   Theoretical Influences

My research was framed by sociocultural theory (Vygotsky, 1978) and informed by perspectives from experiential education (Kolb, 2015) and place-based education (Gruenewald, 2008; Sobel, 2013). Sociocultural theory explains knowledge development as a social process involving the learner and more knowledgeable others, which can include peers, teachers, parents, and community members (Vygotsky, 1978). Experiential education is largely recognized as *learning by doing* where learning is a process of knowledge creation based on experiences and the reflection on and conceptualization of those experiences (Kolb, 2015). Place-based education is a process where learners develop personal relationships with their natural environment or community as a foundation for learning, which can lead to strengthened community bonds, greater appreciation for the environment, and increased engagement as citizens (Sobel, 2013). Gruenewald (2008) explained that education needs to be guided by place-based theory if we expect citizens to be capable of having a positive effect on society and the places we inhabit. Collectively, these perspectives framed the current study because participants constructed knowledge during shared experiences with their peers in the boreal forest within their community.

11.4 Method

In this mixed methods study, the aim was to examine children's activity in the boreal forest by giving them autonomy to explore while using wearable technology. A mixed methods approach was chosen to construct a broader picture of student activity in nature by adding their qualitative interview insights to the largely quantitative GPS data in an explanatory, sequential design (Creswell, 2015). Using a range of methods allowed for triangulation and cross-checking between quantitative and qualitative data (Greig et al., 2012).

11.4.1 School and Outdoor Context

The participating class for this study was selected by first approaching the Superintendent at Yukon Education in Whitehorse, YT. Once the study was explained and permission granted, the next step was to identify a school with a readily accessible wilderness area and meet with its principal. The selected school's location was unique in that despite its small size (i.e., 70 students) it sits on a 2.5-hectare wooded lot in a diverse socioeconomic neighbourhood that adjoins the Chadburn Lake Municipal Park. This park is over 7,000 hectares and consists of several small lakes, untouched stretches of the boreal forest, and Grey Mountain (elevation 1494 m). The total area that the participants explored was approximately 27,000 m^2, of which 60% was forested and 40% was open.

11.4.2 Participants

The principal recommended the Grade 2 teacher with 20 years' teaching experience as possibly willing to participate in several sampling sessions outdoors in all seasons. I met with the teacher to explain my research plan, showed the GPS unit that students would wear, and described how it would collect information important to my research. The teacher agreed to participate, then I invited the class of 15 culturally and linguistically diverse 7- and 8-year-old students to participate. In the initial meeting with the students, I explained the study and described what their role would be; I showed them a GPS unit, modelled how to wear it, and explained how

GPS works using orbiting satellites. Ultimately, 13 students assented to participate although one student did not contribute any data because of repeated absences.

11.4.3 Data Collection and Analysis

Qualitative data were collected through my field notes and interviews I conducted with participating students to gain an understanding of their interest in nature from a participant's perspective over a 4-month period spanning winter and spring. Field notes were taken during the outdoor sessions (Table 11.1); students were interviewed and their responses noted (Table 11.2). I used my field notes to verify the accuracy of my interview notes.

A lexical analysis of the interview notes was conducted by entering the complete dataset into the Free Word Cloud Generator (https://www.freewordcloudgenerator. com/); this resulted in a list of terms with stop words (Rosenberg, 2014) already removed. I refined the list by manually eliminating numbers and proper nouns (e.g., five, [participant names]) and then established word groups based on morphology. For example, zombies (frequency = 9) and zombie (frequency = 4) were merged to become zombies (frequency = 13).

Quantitative data were collected using wearable Garmin eTrex® 10 GPS units (https://www.garmin.com/en-CA/p/87768). These units are relatively small, lightweight, and affixed easily to the wearer's upper arm. Using Garmin Basecamp

Table 11.1 Sample field notes from Session 3 (April 19)	• 1 °C moderate south wind, overcast, 15 cm of snow on ground
	• Using a different material for a GPS strap: stretch-grip tape
	• Heard a loud bang, like a gunshot on the walk to the site; turned out to be an electrical transformer
	• Significant amount of snow gone
	• South-facing slopes are bare, first session with visible bare ground
	• Students again ran right to the pond
	• They found a beaver lodge now that some of the snow is gone
	• Crust of snow supported their weight along pond edge
	• Elizabeth hurt leg on a stick, crying; teacher provided first aid
	• Collecting rocks and sticks
	• Sara decided she wanted her GPS off
	• Most students hanging out at the lodge
	• "Tracking a beaver"
	• "Researching animals"
	• Three digging with sticks in the dirt
	• Run and hide games

Table 11.2 Interview responses from select participants

Maggie	Tao	Maisie	Muhammad
Being chased	Getting things like	Playing games like	Collecting stuff
Seeing a stick that a	moss	Chase the Zombie	Running
beaver ate	I saw beaver tracks	Playing Zombie tag	Seeing lots of spring
Seeing the crocus	Finding things like	Playing with sticks	stuff
flowers	berries	Using our imagination	Berries
Playing Dinosaur	Juniper berries	Playing a bunch of	Climbing a big
Island	Finding kinnikinnick	different games	mountain
Finding crocus flowers	Looking at the loon	Zombies	Some people helped
Putting sticks in the	Playing Zombies	5 knights have hoodies	The Zombie game
pond	Going on the other	game	Seeing lots of animals
Falling into the water	side of the pond	Playing Titanic with	(ducks)
Slipping on the ground	Playing tag	sticks in the pond	Being quiet to see if
Seeing Dakota falling	Playing with my	Playing at the pond	the bird would come
into the pond	friends	The bear	to us
Jumping	Trying to get the bird	Jumping in the water	Seeing a bear
	Running down the hill		Felt happy
	Surprised that Maggie		Seeing bear poop
	got wet		Playing in the woods
	Running down the hill		Seeing Maggie fall
	Playing a game		into the water
	Seeing a bear in the		Watching Dakota
	forest		jump and fall in
	Seeing the beaver, it		Jump challenge
	came close		Finding beavers
	The mud		
	Seeing bird poop, it		
	was red		

software and Google Earth satellite imagery, I overlaid participant movement (e.g., speed and distance) on a topographical image of the landscape to create maps for each participant for each session. Each map was examined for patterns in their movements including where they spent their time, what they avoided, and whether they were together or alone (Fig. 11.1).

Finally, I mixed findings by adding the qualitative themes to each quantitative map to create an infographic as shown in Figs. 11.2–11.5. My interpretations of these infographics were made in accordance with the research question. The steps in my analytical process are presented in Fig. 11.1.

11.4.4 Procedure

In my initial meeting with the Grade 2 teacher, we discussed how we would be careful to not use language that might influence participant movement or decision making unless the participant was in immediate danger. In each of the seven data collection sessions, participants were taken to the same starting point in the neighbouring boreal

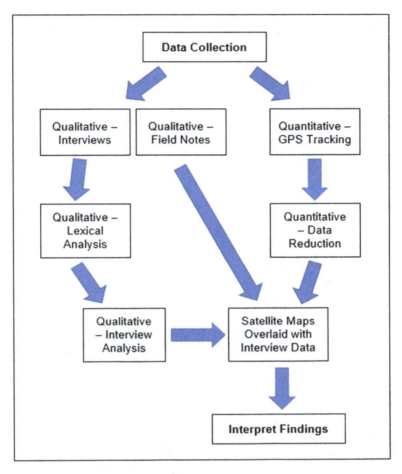

Fig. 11.1 Analytical process

forest, and the GPS units were affixed to them. Then, participants were facilitated through a sensory wake-up circle (Staniforth, 2010), which used the five senses: smell, taste, hearing, touch, and sight. Participants were encouraged to explore their surroundings using their senses as their guide. They were allowed to move freely in the study area for 25–40 min, depending on the session.

During each session, I took field notes on their comments, the games they played, their interactions with nature, the weather including snow depth, and group dynamics (solo vs. group exploration). The verbal command *All in!* was used when it was time to come back to the starting point. I interviewed participants immediately following each session while they sat at their desks and ate snacks. The interviews consisted of the prompt: What did you like most about exploring in nature today? Responses were captured verbatim from 56 interviews across the seven sessions in a notebook.

Fig. 11.2 Maggie's activity in Session 3, as recorded by GPS

Data from Sessions 3 through 7 were the focus of analysis; data from Sessions 1 and 2 were omitted because the novelty of wearing a GPS unit was evident in both the participant interviews and the field notes.

11.5 Results and Discussion

The lexical analysis of the post-session interview responses resulted in a list of words that were then grouped and sorted into themes. Quantitative analysis highlighted the participants' time spent exploring nature, distance travelled, and speed. Mixing of qualitative and quantitative data led to the creation of maps that were overlaid with routes, activities, and themes.

11.5.1 Lexical Analysis

Eleven themes were generated from the lexical analysis: play, actions, fauna, geographic features, flora, people, loose parts, built environment, emotions, sounds, and clothing (Table 11.3). Next, I discuss the five most frequently occurring themes.

Play ($n = 109$) was the most common theme to emerge from participant interviews. It included subthemes of generic play ($n = 59$), specific games ($n = 29$), and running ($n = 21$). Examples of generic play included playing, chasing, and hiding. This type of play is enhanced when students have access to forest and green space (Fjørtoft et al., 2009). Specific games that participants mentioned included Zombies, Dinosaurs, and the Titanic. These types of games are similar to the imaginative and fantasy play that has been observed in other studies of children in nature (Boileau, 2011; Green, 2013, 2016; Waters & Maynard, 2010). Running ($n = 21$), which was included with play because participants typically mentioned it in that context, is an important gross motor activity and a form of risky play. The diverse physical landscape of the study area, which included treeless hillsides, afforded participants the opportunity to run freely. Children engaging in specific types of physical activity afforded by particular

Table 11.3 Results of lexical analysis of post-session participant interviews

Theme	Frequency
Play	109
Actions	88
Fauna	46
Geographic features	35
Flora	20
People	13
Loose parts	8
Built environment	5
Emotions	4
Sounds	4
Clothing	3
Total	**335**

natural surroundings is a finding that reinforces the results of other studies (Fjørtoft et al., 2009; Green, 2013, 2016).

Actions ($n = 88$), the second most frequently occurring theme, included 24 different words/word groups; seeing ($n = 18$), looking ($n = 9$), falling ($n = 7$), and finding, jump, and walking ($n = 6$) occurred most often. This range of actions reflected the diversity of possible interactions with the landscape based on its affordances (Chawla, 2015). The literature suggests that children experience nature through their senses (Boileau, 2011; Green, 2017; Hyun, 2005; Sebba, 1991; Wilson, 1995), which was evident in the present study with the relatively high frequency of seeing and looking.

Fauna ($n = 46$), the third most frequent theme, included mentions of bear ($n = 12$), beaver ($n = 12$), bird ($n = 10$), and poop ($n = 4$). Fauna words and word groups were often referenced in conjunction with sensory actions, for example, "seeing the bear," "looking at the loon," and "seeing lots of animals." This theme is consistent with the results of other recent studies involving children and nature (Boileau, 2011; Green, 2016; Kalvaitis & Monhardt, 2012; Waters & Maynard, 2010).

Geographic features ($n = 35$), the fourth most frequent theme, included hill ($n = 8$), pond ($n = 7$), water ($n = 7$), and forest ($n = 4$). The pond was the geographic feature that participants mentioned most often because they used other words (e.g., lake, water) to describe it. The pond was sometimes mentioned in conjunction with play, for example, "playing at the pond" and "jumping in the water;" however, its presence also led to mentions of fauna, for example, "the beaver slapped its tail" and "looking at the duck." The hill geographic feature reinforces the affordance of natural spaces for physical activity, for example, e.g., "running down the hill." Diversity in outdoor landscapes has been linked to enhanced play and physical activity in children (Fjørtoft et al., 2009).

Flora ($n = 20$), the fifth most frequent theme, included trees ($n = 4$), berries ($n = 3$), and specific types of plants ($n = 7$; e.g., daisy, kinnikinnick). Words in the flora theme were often associated with senses (e.g., "seeing the crocus flowers") or were sometimes simply identification (e.g., "juniper berries"). Flora were a major attraction; all participants mentioned flora at some point during their interview. This attraction may reveal an opportunity for using place as a focal point for teaching and learning (Sobel, 2013) and for building upon student interest, which could have positive effects on student motivation and agency (Ghafouri, 2014).

11.5.2   Quantitative Analysis

The quantitative data collected using the GPS units offered insight into participants' autonomous movement in nature. These data were in the form of satellite maps that displayed where participants went and showed the features of that terrain. Here I highlight the GPS maps from four participants whom I purposefully selected because they broadly represent the overall group (Table 11.4). For example, if a participant

Table 11.4 Results of GPS data for selected participants in Session 3

Participant	Time in forest (%)	Time in open space (%)	Distance travelled (m)	Elapsed time (min)	Average speed (km/h)
Maggie	54	46	569	38:09	1.0
Tao	62	38	421	36:27	0.7
Maisie	64	36	688	32:48	1.3
Muhammed	54	46	507	38:31	0.8

Note All names are pseudonyms

typically explored with a particular friend, then only one of the two participants was selected. Variables such as time spent exploring, distance travelled, and average speed were also considered.

11.5.3   Mixed Analysis

The maps with the GPS data were overlaid with the qualitative data collected through interviews to provide a mixed representation of the students' autonomous experiences in the outdoors. In the following sections, I reconstruct and examine these experiences for each of the four representative participants.

11.5.3.1   Maggie

Maggie represents participants who were typically involved in play and other actions while out in nature but who were occasionally engaged by surrounding flora and fauna. For example, after three of the five sessions, Maggie commented about the flora and fauna that she observed during her explorations: "seeing a stick that a beaver ate" and "finding crocus flowers." Aspects of ecology are a consistent draw for children when they are given autonomy to explore the outdoors (Ghafouri, 2014; Green, 2016). Figure 11.2 shows Maggie's autonomous travels across the landscape during Session 3. Across all five sessions, Maggie spent an average of 63% of the time in the forest and 37% of the time in clearings. This result supports findings from other studies, where children chose to spend more time in forested areas (Fjørtoft et al., 2009). The distance that Maggie travelled each session ranged from 336 to 963 m, which can be explained by her actions in and interactions with nature. In the session when Maggie travelled 336 m, her comments revealed that they played a game called Dinosaur Island and spent time putting sticks in the pond, which explains the shorter distance travelled. In the session when Maggie travelled 963 m, she commented about seeing crocus flowers; the GPS data made it clear that she spent time searching across the hillside where crocuses typically grew. These connections found by correlating distance travelled with interview data reflect place-based learning and sociocultural theories. An element of place, in this case a spring flower, influenced Maggie's experience (Sobel, 2013). Maggie played the co-constructed game Dinosaur Island, which was an example of sociocultural processes when interacting with peers (Vygotsky, 1978).

11.5.3.2   Tao

Tao represents participants who were focused on flora and fauna in every session but who also commented on engaging in generic play, specific games, and running during some sessions. In each session, Tao typically commented on two aspects of flora and fauna that he had noticed, for example, "finding things like berries," "looking at the loon," and "seeing a bear." The data for Tao showed diverse travels in each session, which might be explained by his search for interesting aspects of flora and fauna (Boileau, 2011; Kalvaitis & Monhardt, 2012). Figure 11.3 shows Tao's explorations in nature during Session 3, when 62% of the time was spent in the forest and 38% in more open spaces. On average, Tao spent 66% of his time in the forest and 34% in clearings. Tao was action oriented and commented on "finding things like berries," "going on the other side of the pond," and "running down the hill." Tao's movements during his seemingly tireless search for artifacts of flora and fauna are an important example of place-based learning theory where the focus of attention is on the landscape and its affordances (Sobel, 2013).

11.5.3.3   Maisie

Maisie represents participants who spent most (i.e., 70%) of their time in forested areas and who spent time most sessions playing, mentioning games in all but one interview. GPS data revealed that Maisie went directly to a dense area of forest next to the pond during each session (Fig. 11.4). Maisie's interview comments offered insights into her travel patterns, for example, "playing at the pond", "playing with sticks" and "playing Titanic with sticks in the pond". This finding is consistent with other studies where children's activities were influenced by what nature afforded (Fjørtoft et al., 2009; Green, 2013, 2016). Maisie was not playing alone; other participants also chose to play games in this densely forested area, which afforded multiple hiding spots. Maisie's repeated references to play or playing are consistent with other studies involving children's explorations in nature (Boileau, 2011; Green, 2013; Kalvaitis & Monhardt, 2012). Elements of sociocultural theory may help to explain Maisie's movements in nature because they tended to engage in play with peers (Vygotsky, 1978).

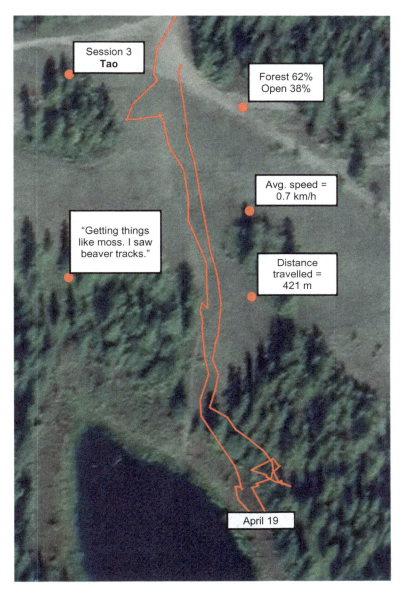

Fig. 11.3 Tao's activity in Session 3, as recorded by GPS

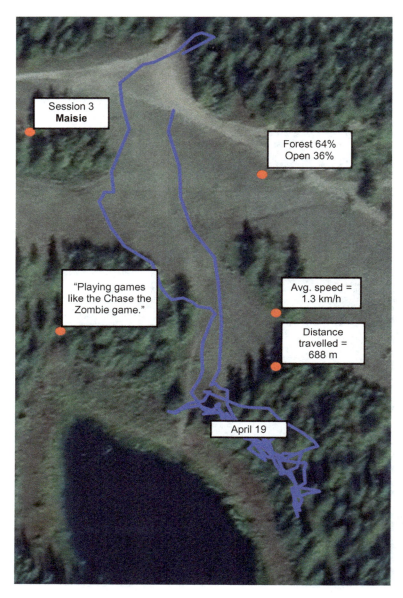

Fig. 11.4 Maisie's activity in Session 3, as recorded by GPS

Fig. 11.5 Muhammad's activity in Session 3, as recorded by GPS

11.5.3.4 Muhammad

Muhammad represents participants who did not have a particular focus for their explorations, choosing to spend time playing, engaging in actions, or observing flora and fauna depending on the session. For example, Muhammad commented about play during three sessions, about actions such as jumping and climbing during all five sessions, and about flora and fauna in four sessions. Muhammad's comments after Session 3 (Fig. 11.5) clearly demonstrate this eclectic behavior. The GPS data that showed him zigzagging across the landscape might be explained by his openness to engage in multiple affordances of the natural environment. This openness might also explain why he spent 61% of the time in the forest and 40% in open spaces, which was the most time spent in clearings of all the participants. Children have been found to be drawn to a diversity of places in nature rather than fixating on a single location (Green, 2013). Muhammad's movement exhibited a degree of spatial autonomy in nature that has been linked to a child's confidence in their surroundings (Green, 2017) and a willingness to engage in risky play (Sandseter, 2007).

11.5.4   Summary of Results and Discussion

The mixed data for Maggie, Tao, Maisie, and Muhammad provide an overview of their authentic interests in nature, giving a clear picture of where they travelled and what they did when given autonomy. This overview reflects the choices and interests of the entire group of participants who were also engaged in play, actions, and observations that were shaped by the landscape. Like the four representative participants, the larger group was involved in generic play (e.g., "playing with my friends") and in specific games involving their imagination and fantasy (e.g., Zombie Elimination, Dinosaur Island, Titanic) shaped by the affordances of the landscape. The larger group also engaged in a wide range of actions supported by geographic features or involving the flora and fauna that were part of the landscape (e.g., "jumping around the mud," "finding kinnikinnick," "throwing sticks in the water," "trying to get the bird to come over," "going in trees and hiding," "looking at four squirrels"). Flora and fauna were a focus for several of the representative participants, and these items represented 20% of the words and word groups mentioned in all interviews. Across the five sessions, all participants demonstrated autonomy as they made specific choices about where they went and what they did.

11.6   Concluding Remarks

In this study, I sought to uncover 7- and 8-year-old children's authentic behaviours in the boreal forest over a 4-month period from winter to spring. Specifically, I investigated where participants chose to go when given autonomy in a natural landscape and what they chose to do. Other studies have shown that adult presence can affect children's behaviours (e.g., Green, 2016); however, in this study geo-spatial technology was employed to track their authentic and autonomous movement in nature in a minimally invasive manner. The technology itself led to a few minor inconveniences; for example, if the GPS units got too cold, they would experience battery failure that, on a few occasions, required replacement mid-session. Additionally, participants who played vigorously in dense forest would occasionally knock the GPS loose.

To supplement the GPS data and gain a deeper understanding of their autonomy, I asked the participants about their experiences in nature. However, a limitation of this study was that I asked a single question in a group setting and it is possible that some were influenced by the responses of others near them. Solutions would include asking additional questions to elicit richer responses and conducting individual interviews in a space away from the group. I focused on a small group of participants from a single class, which could be considered a limitation, and future research could include a wider range of participants (e.g., age, location).

I obtained findings by cross-referencing participants' quantitative movement data with their qualitative interview data and triangulating with my field notes. These findings reinforce the results of other studies of children in nature and extend them by focusing on participants in a northern region of Canada that is often overlooked in the literature. I found that participants were drawn to the natural landscape, which other researchers have noted offers a complex and dynamic setting when compared to manufactured playgrounds and human-built schoolgrounds (Waters & Maynard, 2010). The natural landscape afforded a learning environment that these children appeared to be drawn to, which reflects previous studies (e.g., Fjørtoft et al., 2009; Green, 2013, 2016; Waters & Maynard, 2010). For example, forests have an abundance of loose parts such as sticks, leaves, moss, and cones that may contribute to children's preferences for these spaces (Boileau, 2011; Green, 2013; Waters & Maynard, 2010); loose parts were a focus for my participants. Young students should be provided with opportunities to play and interact with loose parts, particularly in a natural environment, to promote imagination and peer interaction. A simple stick can become many different things when a child is engaged in socially constructed play with peers.

Access to natural spaces with abundant loose parts for early childhood and primary school education should be prioritized by school and district administrators. Although many Yukon schoolgrounds are close to the boreal forest, most have fixed play structures and lack natural artifacts. Schoolgrounds could be populated with loose parts that reflect the elements of the natural landscape. Recognizing the importance of children interacting with loose parts and in a natural setting could lead to the design—or redesign—of schoolgrounds. Risky play, a frequent choice of my participants that was supported by the geographic features of the natural landscape, should be embraced despite the overly restrictive policies about accessing the outdoor learning environment that sometimes exist. Risks associated with the natural environment can be given an equal level of acceptance as the hazards associated with indoor learning environments. My findings make a strong case for embracing authentic student interest as a springboard to authentic inquiry. The ability to build curriculum around an existing area of student interest would be an advantageous position for any school system.

References

Boileau, E. Y. S. (2011). *"It's alive!": An exploration of young children's perceptions of the natural world* (Publication No. MR84622). Master's thesis, Royal Roads University (Canada). ProQuest Dissertations and Theses Global.

British Columbia Ministry of Education. (2022). *Curriculum orientation guide.* https://curriculum.gov.bc.ca/curriculum/science

Caiman, C., & Lundegård, I. (2014). Pre-school children's agency in learning for sustainable development. *Environmental Education Research, 20*(4), 437–459. https://doi.org/10.1080/13504622.2013.812722

Chawla, L. (2015). Benefits of nature contact for children. *Journal of Planning Literature, 30*(4), 433–452. https://doi.org/10.1177/0885412215595441

Chawla, L. (2020). Childhood nature connection and constructive hope: A review of research on connecting with nature and coping with environmental loss. *People and Nature, 2*(3), 619–642.

Cheng, J.C.-H., & Monroe, M. C. (2012). Connection to nature: Children's affective attitude toward nature. *Environment and Behavior, 44*(1), 31–49. https://doi.org/10.1177/0013916510385082

Creswell, J. W. (2015). *Educational research: Planning, conducting, and evaluating quantitative and qualitative research* (5th ed.). Pearson.

Deci, E. L., & Ryan, R. M. (2000). The "what" and "why" of goal pursuits: Human needs and the self-determination of behavior. *Psychological Inquiry, 11*(4), 227–268.

Ernst, J. (2014). Early childhood educators' use of natural outdoor settings as learning environments: An exploratory study of beliefs, practices, and barriers. *Environmental Education Research, 20*(6), 735–752. https://doi.org/10.1080/13504622.2013.833596

Ernst, J., & Tornabene, L. (2012). Preservice early childhood educators' perceptions of outdoor settings as learning environments. *Environmental Education Research, 18*(5), 643–664. https://doi.org/10.1080/13504622.2011.640749

Fjørtoft, I., Kristoffersen, B., & Sageie, J. (2009). Children in schoolyards: Tracking movement patterns and physical activity in schoolyards using global positioning system and heart rate monitoring. *Landscape and Urban Planning, 93*(3), 210–217. https://doi.org/10.1016/j.landurbplan.2009.07.008

Ghafouri, F. (2014). Close encounters with nature in an urban kindergarten: A study of learners' inquiry and experience. *Education 3–13, 42*(1), 54–76. https://doi.org/10.1080/03004279.2011.642400

Green, C. (2013). A sense of autonomy in young children's special places. *International Journal of Early Childhood Environmental Education, 1*(1), 8–31. https://naturalstart.org/research/ijecee/volume-1-number-1

Green, C. (2016). Sensory tours as a method for engaging children as active researchers: Exploring the use of wearable cameras in early childhood research. *International Journal of Early Childhood, 48*(3), 277–294. https://doi.org/10.1007/s13158-016-0173-1

Green, C. (2017). Children environmental identity development in an Alaska native rural context. *International Journal of Early Childhood, 49*(3), 303–319. https://doi.org/10.1007/s13158-017-0204-6

Greig, A., Taylor, J., & MacKay, T. (2012). *Doing research with children* (2nd ed.). SAGE.

Gruenewald, D. A. (2008). The best of both worlds: A critical pedagogy of place. *Environmental Education Research, 14*(3), 308–324. https://doi.org/10.1080/13504620802193572

Hughes, J., Rogerson, M., Barton, J., & Bragg, R. (2019). Age and connection to nature: When is engagement critical? *Frontiers in Ecology and the Environment, 17*(5), 265–269.

Hyun, E. (2005). How is young children's intellectual culture of perceiving nature different from adults'? *Environmental Education Research, 11*(2), 199–214. https://doi.org/10.1080/13504620420000338360

Kahn, P. H., Jr., Severson, R. L., & Ruckert, J. H. (2009). The human relation with nature and technological nature. *Current Directions in Psychological Science, 18*(1), 37–42. https://doi.org/10.1111/j.1467-8721.2009.01602.x

Kahn, P. H., Weiss, T., & Harrington, K. (2020). Child-nature interaction in a forest preschool. In A. Cutter-Mackenzie-Knowles, K. Malone, & E. Barratt Hacking (Eds.), *Research handbook on childhoodnature* (pp. XX–XX). Springer. https://doi.org/10.1007/978-3-319-67286-1_33

Kalvaitis, D., & Monhardt, R. M. (2012). The architecture of children's relationships with nature: A phenomenographic investigation seen through drawings and written narratives of elementary students. *Environmental Education Research, 18*(2), 209–227. https://doi.org/10.1080/13504622.2011.598227

Kolb, D. A. (2015). *Experiential learning: Experience as the source of learning and development* (2nd ed.). Prentice Hall.

Loebach, J., & Gilliland, J. (2010). Child-led tours to uncover children's perceptions and use of neighborhood environments. *Children, Youth and Environments, 20*(1), 52–90. https://doi.org/10.7721/chilyoutenvi.20.1.0052

Louv, R. (2005) *Last child in the woods: Saving our children from nature-deficit disorder*. Algonquin Books.

Prince, H., Allin, L., Sandseter, E. B. H., & Ärlemalm-Hagsér, E. (2013). Outdoor play and learning in early childhood from different cultural perspectives. *Journal of Adventure Education & Outdoor Learning, 13*(3), 183–188.

Rickinson, M., Dillon, J., Teamy, K., Morris, M., Choi, M.-Y., Sanders, D., & Benefield, P. (2004). *A review of research on outdoor learning*. National Foundation for Education Research and Field Studies Council.

Rosenberg, D. (2014). Stop, words. *Representations, 127*(1), 83–92. https://doi.org/10.1525/rep.2014.127.1.83

Rosenow, N., & Bailie, P. (2014). Greening early childhood education [Special issue]. *Children, Youth and Environments, 24*(2), 1–9. https://doi.org/10.7721/chilyoutenvi.24.2.0001

Sandseter, E. B. H. (2007). Categorising risky play—How can we identify risk-taking in children's play? *European Early Childhood Education Research Journal, 15*(2), 237–252. https://doi.org/10.1080/13502930701321733

Sebba, R. (1991). The landscapes of childhood: The reflection of childhood's environment in adult memories and in children's attitudes. *Environment and Behavior, 23*(4), 395–422. https://doi.org/10.1177/0013916591234001

Sobel, D. (2013). *Place-based education: Connecting classrooms & communities* (2nd ed.). Orion Society.

Staniforth, S. (2010). *Get outdoors: An educator's guide to outdoor classrooms in parks, playgrounds and other special places*. Retrieved from https://www.hctfeducation.ca/file/get-outdoors.pdf

Vygotsky, L. S. (1978). *Mind in society: The development of higher psychological processes*. Harvard University Press.

Waters, J., & Maynard, T. (2010). What's so interesting outside? A study of child-initiated interaction with teachers in the natural outdoor environment. *European Early Childhood Education Research Journal, 18*(4), 473–483. https://doi.org/10.1080/1350293X.2010.525939

Wilson, R. A. (1995). Nature and young children: A natural connection. *Young Children, 50*(6), 4–11. http://www.jstor.org/stable/42727088

Jesse Jewell BAH (Nipissing University), BEd (Nipissing University), MA (University of Victoria), is principal and education consultant at Yukon Wild School, a K-7 nature-based school. Jesse previously worked as an experiential education consultant serving K-12 Yukon School for over a decade. As a classroom science and geography teacher, Jesse taught everywhere but the classroom, preferring the boreal forest. Jesse's research interest is focused on the student-nature relationship and how autonomy enables more authentic insight into student interest.

Todd M. Milford BSc (University of Victoria), BEd (University of Victoria), DipSpecEd (University of British Columbia), MEd (University of Victoria), PhD (University of Victoria), is an associate professor and chair in the department of Curriculum and Instruction at the University of Victoria. Prior to this he was a lecturer in the Art, Law, and Education Group at Griffith University in Brisbane Australia. He has science and special education classroom teaching experience as well as in the online environment. He has been teaching at the postsecondary level since 2005 primarily in the areas of science education, mathematics education, and classroom assessment. His research has been and continues to be varied; however, the constant theme is using data and data analysis to help teachers and students in the classroom.

Christine D. Tippett BASc (University of British Columbia), BEd (University of Victoria), MA (University of Victoria), PhD (University of Victoria), is an associate professor of science education in the Faculty of Education at the University of Ottawa. She was an engineer before she obtained her teaching degree, which influences her ways of thinking about science education. Her research interests include visual representations, science education for all students, and professional development for science educators (preservice, inservice, and informal). Current projects focus on preservice science teachers' images of engineers, early childhood STEM education, and assessment of representational competence.

Chapter 12
Nature Is Our Classroom: Place-Conscious Pedagogy and Elementary Science Education

Sharon Pelech and Darron Kelly

12.1 Introduction

How do we connect science education to the lives of students and to the contexts in which they live? This chapter explores the natural connections between place-conscious pedagogy and science education in an elementary school setting. This exploration involves a case study in a K–6 school in Edmonton, Alberta. The school is nestled beside an urban forest and river system and is surrounded by newly developed suburbs that have replaced once-productive farmland. In this chapter, we show that place-conscious pedagogy provides authentic learning opportunities and, perhaps more importantly, helps students gain a sense of agency and meaningful connection to their lived environment.

Following from this study, we conclude that teachers who bring a sense of place consciousness to their teaching can create genuine student interest in science by drawing on children's natural curiosity and wonderment for the world around them. This approach to teaching goes beyond a focus on engagement to provide students with opportunities to develop emotional affinities that can lead to environmental stewardship and caring for their community, what Ormond (2013) termed place attachment. As we have observed from research in other settings, students form a greater "sense of kinship, belonging and commitment to place while extending their understanding of the interconnectedness of all places" (Kelly & Pelech, 2019a, p. 738).

S. Pelech
University of Lethbridge, Lethbridge, AB T1K 1N5, Canada
e-mail: sharon.pelech@uleth.ca

D. Kelly (✉)
Memorial University, St. John's, NL 1B 2B4, Canada
e-mail: p45dik@mun.ca

C. D. Tippett and T. M. Milford (eds.), *Exploring Elementary Science Teaching and Learning in Canada*, Contemporary Trends and Issues in Science Education 53,
https://doi.org/10.1007/978-3-031-23936-6_12

Strengthening the link between place-conscious pedagogy and science education supports recent scholarship that recognizes addressing "unprecedented issues of
global climate change, food sovereignty, diminishing fresh water resources and the
like" will require new and multiple perspectives and practices in education (Avery &
Hains, 2017, p. 130). We view the diversity of projects, practices, and directions
that teachers bring to place-conscious pedagogy (PCP) by exercising their agency
and local knowledge as key to addressing these issues. In this chapter we explore
how, driven by the conviction and expertise of teachers, PCP can encourage children
to develop the essential attributes of natural inquirers and problem solvers who can
achieve provincial curriculum objectives and successfully confront the most urgent
concern of our time, environmental sustainability (Alberta Education, 1996, 2017).
This chapter explores the natural connection of PCP and science education and then
provides a case study that demonstrates what is possible when a teacher allows the
natural environment to inspire students to become engaged as science learners.

12.2 Place-Conscious Pedagogy

Similar to the traditions of place-based education and environmental education,
place-conscious pedagogy encourages rich, meaningful connections between schools
and their social and environmental settings by contextualizing school subjects within
the local (Chang, 2017; Gruenewald, 2003; Kelly, 2005, 2006, 2007; Kelly & Pelech,
2019b; Lescure & Yaman, 2014; Sedawi et al., 2021). Sobel (1996, 2005) argued that
school structures and ways of implementing curriculum often do not include the local
environment and, as a result, "devalue local cultural identity, traditions, and history in
preference to a flashily marketed homogeneity" (p. 1). As we will describe in more
detail below, PCP is an authentic approach to teaching by recognizing the educational value of place and student agency as the context from which to bring science
curriculum to life (Gruenewald, 2003; Gruenewald & Smith, 2010). Examples of
place-based initiatives include establishing a community garden, building an outdoor
classroom, and observing the impact of human activity on local ecosystems (Collyer,
1998; Gruenewald & Smith, 2010; Kelly, 2005). The impact of place-conscious
teaching for improved academic performance and engagement has been well documented (Harvard Graduate School of Education, 1999a, 1999b; Liberman & Hoody,
1998; Nichols et al., 2016; Smith, 2007); but this approach can also move beyond
academic achievement to strengthen student affinity to the community and the natural
world, supporting the development of informed, active, engaged citizens (Deringer,
2017; Kelly & Pelech, 2019a; Smith, 2002; Sobel, 2005).

Our conceptualization of PCP expands beyond the traditional aspects of place-
based education to include a diverse understanding of place and a critical interest
in enhancing transformative agency. We pursue our expanded interest through case
studies of place-conscious teaching practices and theory-driven thematic analysis.

Along with the experiential, hands-on curricular learning that can occur within
community and environmental settings, PCP recognizes the effect of situating

curriculum within local culture and ecology for developing "relations between humans and other animals, plants, and their habitats" (Avery & Hains, 2017, p. 130). We see these relationships as crucial for addressing the many of the environmental crisis we are currently experiencing locally and globally (some examples include global warming, deforestation, massive wildfires, flooding, species loss).

There are numerous critiques of the taken-for-granted meaning of *place* as it appears in the environmental and science education literature (e.g., Avery & Hains, 2017; Coughlin & Kirch, 2010; Greenwood, 2013; Gruenewald, 2003, 2008; Semken & Butler Freeman, 2008). van Eijck and Roth (2010) argued that the ideological discourse of place-based education often assumes place is one ideal version of the natural world, stating:

> Place is not simply a location that we can identify by listening to a particular voice. It is a location unfolding in time just because people inhabit, visit, rebuild, make, enjoy, sorrow, describe, and recount, hence live it, by which it is articulated by a multitude of voices. (p. 882)

In other words, from a critical pedagogical framework, place is a social construct informed by the varied history, culture, and assumptions of the people that inhabit the place.

We have previously argued that through a pedagogy of place it is crucial that students are provided with an opportunity and means of "analyzing, critiquing, and improving local assumptions and practices" (Kelly & Pelech, 2019a, p. 734). By ignoring the complex multifariousness of how place is experienced and understood, critical consideration of local issues such as habitat loss, dysfunctional community relationships (both human and human to non-human), and patterns of consumption that continue without discussion of sustainability remain "unidentified, unchallenged, and unchanged" (Kelly & Pelech, 2019a, p. 733). Similarly, Pelo (2014) argued that if we ignore the ecological identity of the local, we lose the "intimate connection to the land, the sky, the air. Any place can become home, we're told. Which means, really, that no place is home" (p. 43). Therefore, no one questions why food is shipped from other countries, when farmland is converted to suburbs, rivers are dammed, and forests are bulldozed for roads. In light of Pelo's argument, PCP asserts that place is a "living ecological relationship" (Kudryavtsev et al., 2012, p. 231) that includes multiple voices and focuses on the importance of being able to explore and learn from the "symbiotic relationships between schools and their communities" (Lynch et al., 2017, p. 711). PCP can offer students an ability to engage in authentic environmental science through the identification of critical issues and the posing of vital questions. Asfeldt et al. (2009) contended that once children are connected to their local place similar connections will expand children's understanding outward to encompass other places, so that care and concern for our global place will follow.

A benefit of PCP is its natural connection to enhanced student agency. Biesta and Tedder (2007) defined agency as a "capacity for autonomous social action or the ability to operate independently of determining constraints of social structure" (p. 135). However, interpretations of student agency often focus on a narrow view of motivation and engagement in order to achieve predetermined learning outcomes (Goodman & Eren, 2013) instead of addressing student self-determination and their ability to find meaning in learning and effect change in their social and natural environments. Our research explores teaching practices that demonstrate how students can actively participate in understanding and shaping the world around them and, as a result, experience the capability they have as agents of change within their community (Basu & Barton, 2010; Kelly & Pelech, 2019b). Our work shows a transformative experience of agency can emerge for students by situating curriculum in and by creating direct connections with place. Students' sense of agency develops as teachers provide local contextual learning opportunities and invite students to rethink the place they live in, leading to students' increased ability to intervene and influence their local communities and learning environments (Pelech & Kelly, 2020).

12.3   Science Education and Connection to Place

Think for a moment of a science textbook that is describing a food chain. For most teachers, the sample food chain is something like:

$$\text{grass/seeds} \rightarrow \text{rabbit} \rightarrow \text{coyote}$$

This food chain example is often represented in textbooks throughout Canada (viz., *Science in Action 7*, Addison Wesley, 2001, pp. 35–38). But where does this food chain actually exist? What does this food chain look like in southern Alberta compared to Newfoundland and Labrador? Where are humans located within this food chain?

As a result of the generic examples and topics that are often found in science textbooks and curriculum guides, scientific knowledge is often represented as isolated topics, fragmented and separate from each other and from specific places. Scientific concepts are seen as a collection of individual outcomes that need to be covered, often removed from the greater, complex contexts where they actually operate (Blades, 2001; Dewey, 1902; Eger, 1993; Kalas & Raisinghani, 2019; Lyons, 2006; Pelech, 2015). Bigelow (2014) argued that his schooling suppressed connections of the academic subjects from a sense of place:

> Through silence about the Earth and the Indigenous people.... We actively learned to *not think* about the Earth, about the place where we were. We could have been anywhere – or nowhere.... School erected a Berlin Wall between academics and the rest of our lives. (p. 37)

PCP offers an opportunity to actively and deliberately counter generic, fragmented facts by connecting science back into the world. By connecting scientific understandings like food chains and food webs within local contexts, students can connect such concepts to their environment and deepen their appreciation of the place where they live. As students are introduced to local ecosystems and their integration within these complex systems, their understanding of ecological concepts becomes more nuanced and supportive of a clear sense of social and environmental responsibility (Kalas & Raisinghani, 2019; Semken & García, 2021). Kudryavtsev et al. (2012)

proposed that what they call place "rootedness [leads to a] sense of deep care and concern for that place" (p. 233) as a motive for further engagement and action.

PCP can support science education as a diverse and expansive form of teaching that includes multiple ways of entering and understanding relationships within the natural environment. It offers an opportunity to acknowledge connections with place and multiple ways of knowing that traditional science classrooms often ignore. Eger (1993) argued that science classrooms often suppress the aspects of science that are embedded in histories, philosophies, and sociologies by focusing on simply teaching facts and problem-solving skills:

> [Science] takes on a cut-and-dried appearance, a learn-it-and-use-it format, and an aura which announces that nothing profound, mysterious or specifically human is to be found here…. For if science is both distasteful and unenlightening (about the world), its value can lie only in professional ambition. (p. 4)

As a result, science education is often mischaracterized as the memorization of content or the application of techniques or calculation methods that are disconnected from the deeper traditions that gave rise to them (Crease, 1997; Pelech, 2015). PCP offers opportunities to embrace diversity in multiple epistemological stances and values in science. By including Indigenous ways of knowing (Bang & Medin, 2010; Skilbeck & Stickney, 2020) and local cultural knowledge (Avery & Hains, 2017; Üztemur & Dere, 2022), PCP resists the pedagogy of erasure that often happens throughout the process of standardization of science curriculum, disintegration of techniques and procedures, and an inadvertent fragmentation and dissociation of scientific concepts.

What is often lost when presenting science as predetermined bits of knowledge is the tentativeness and creativity required for authentic scientific understanding and dispositions to occur (Aikenhead, 2003). As a result, students experience science education as a "contrived experience, understanding for memorization" (Blades, 2001, p. 70) that is focused on their ability to recall information for writing and passing examinations, as opposed to one of many diverse ways of understanding and relating to the world around them (Avery & Hains, 2017; Bang & Medin, 2010; Blades, 1997; van Eijck & Roth, 2010; Wong & Hodson, 2010). PCP offers students an opportunity to experience the tentativeness and creativity—beginning with authentic questions that come from experiences in place and working through the process—in coming to understand how to best address the question through the scientific process.

12.4 Framing the Research

This case study is part of a current, larger research project that is documenting the experiences of teachers who self-identify as practitioners of place-conscious peda-gogy in Alberta and in Newfoundland and Labrador. The participating teachers in the larger project are very generous with their time and ideas and enthusiastic about sharing their work. Our main research method is semistructured, open-ended interviewing (Babbie & Benaquisto, 2010; Seidman, 2019). Along with the interviews, we collect artifacts and take digital photographs to document the place-conscious projects, practices, and student works that result from this approach to teaching (Creswell & Miller, 2000; Schensul et al., 1999).

From the rich data, ideas, and practices provided by participating teachers, we conducted an analysis (Seidman, 2019) that relates our central themes to PCP and provides knowledge of PCP's overall value to science education. As key themes are identified and developed, we note the unique and diverse ways in which PCP is enacted in these two provinces. Piersol (2013) observed that there is no "magical prescription [for authentic teaching that can be prescribed across all settings; rather each case is inherently contextual, attuned to the location, and demonstrates the pedagogical agency of teachers who step outside to] listen and dig into the stories of place" (p. 64). Yet each of these unique situations—each case—demonstrates possibilities for high-quality curricular learning and active pathways for students to achieve meaningful social and environmental transformation.

The case study described here draws from this rich data set and illustrates a truly remarkable elementary school and its teachers who embraced place-consciousness as the central tenet of their teaching. The school offers a unique situation for the study of PCP in relation to science education as the majority of teachers and administrators work to incorporate this idea into daily life and practice. The teachers integrate cultural, natural, and historical aspects of place into every subject area and invite students to actively engage, challenge, and expand their understanding of the local. In applying this approach to elementary science education, the teachers foreground the natural interconnections the school community has to the local environment and encourage students to explore and discover the human and more-than-human relationships around them (Gruenewald, 2003).

12.5   Setting the Stage

Roberta MacAdams School is a K–6 public school that opened in 2016 on the south side of Edmonton, Alberta. Edmonton was established on what was a gathering place for many Indigenous peoples before European settlement. The city is situated on Treaty 6 territory—the traditional lands of "the Cree, Blackfoot, Métis, Nakota Sioux, Iroquois, Dene, Ojibway/Saluteaux/Anishinaabe, Inuit, and many others" (University of Alberta, 2021). The school was built to support a relatively new subdivision that borders an urban forest and Blackmud Creek. As one enters the school, the connections to the land, the community, and the local Indigenous peoples are evident. From a teepee (raised in ceremony with a local Elder) prominently placed in the library to the books and artifacts on display, there is a vibrant sense of school community. Nearby display cases are filled with books and items from nature. The school-vision statement *Let nature be your teacher* and the school motto—*Every child, every day*—are printed in bold letters on a wall in the light-filled foyer. On any given day of the school year, students can be seen outside in the schoolyard, at the nearby pond, or making their way to and from the nearby forest. In short, the school is an ideal site to study the educational value of place-conscious pedagogy. The case study in this chapter documents several teaching initiatives that demonstrate how one class embraced place consciousness and its effects on science education.

12.6   Let Nature Be Your Teacher

The main data source for the case study is an in-depth interview conducted with Meghan, a dynamic teacher with over 16 years of experience teaching students from Grades 3 to 8. Meghan was one of the initial teachers hired when Roberta MacAdams School first opened; she has taught Grade 4 for the past three years. Throughout the interview, Meghan described many different projects related to PCP and spoke passionately of the creativity and enthusiasm that both she and her students experienced throughout these projects. When asked what drew her to place-conscious teaching, Meghan explained how she has always taught using inquiry but, when she came to Roberta MacAdams School and saw the surroundings, she became excited about the possibilities for local learning. The staff began the very first year of the school by going together on a nature walk. From that initial, personal experience, Meghan rekindled a sense of wonder about the world around her; she decided to try and provide the same kind of experiences for her students. Karen the school principal, also participated in this interview. Under Karen's leadership, the teachers in the school are committed to engaging students in PCP throughout the grades. The collaborative work that the school staff are involved in has created the needed space for teachers and students to connect to the local community and environment throughout the school year.

A key element that Meghan recognized as essential for authentic learning was to be flexible and open to what emerged through students' interactions with nature. Meghan said that "nature always [brings] something to us. Sometimes I go out with certain intentions but something else magnificent would happen. I knew that I had to allow it to unfold." The willingness and trust to create space for something to happen that will catch the curiosity and wonder of students can be challenged in the current climate of predetermined curricular outcomes and standardized testing—priorities that are especially prominent for science education in many jurisdictions. Yet, when this space is created for students to experience their world and wonder about it, rich and meaningful learning opportunities can begin to flourish. As Piersol (2013) observed:

> The land is teaming with signs; whether they are tracks of coyote or that of an old fence post, these clues weave us into the stories and spirit of place. It may take time for the stories to unfold.… I have found that children with their keen eyes and excitement for details … help us to rediscover the extraordinary in the ordinary. (p. 68)

Although the flexibility and openness to trust in the curiosity of children can be difficult for teachers accustomed to a system that focuses on canonical science knowledge to prepare students for high-stakes examination and postsecondary education (Tytler, 2010), the understanding of scientific concepts and processes available through engaging in place is unique. It goes beyond basic knowledge to what Pirrie and Gillies (2012) described as essential elements of science as a discipline, including multiple interpretations of phenomena and a working knowledge of concepts seen in the world.

When asked to give an example of when something spontaneous happened on a nature walk with her class, Meghan gave many instances. One instance was when the class went to look for the beaver at the nearby pond; however, while walking the path, students came upon an intact wasp nest on the ground but still attached to a branch. Meghan and the students began to discuss how to look for signs that the nest may still be active and, after determining it was safe, brought it back to the classroom. From this impromptu discovery, many questions and wonderments emerged as the students' curiosity began to peak. Students generated a wide variety of questions about the difference between bees and wasps, how they lived and how they built nests, which quickly grew into a chorus of questions the students wanted to answer. Piersol quoted Sobel (2008) that "when students get really enraptured in a topic and start to search for pieces of information, see the connections between different ideas, and then glimpse the big pattern, they're really engaged in a kind of treasure hunt" (p. 55). Piersol (2013) described these treasure hunts as having "the goal of sharing the unique natural and cultural heritage of an area" (p. 67). This disposition for discovery is at the heart of science, and its development can be nurtured when students connect their curiosity to place and generate their own meaningful questions. From the initial finding and wondering about the wasp nest, Meghan was able to extend the student inquiry by finding experts from the community who brought beehives to the school and talked with the students about bees and wasps, how bees make honey, and why it is important to protect bees for pollination. This example illustrates how meaningful questions that emerge from student encounters with place are fueled by wonderment and curiosity. Encounters with place, in turn, lead to deeper explorations of the science within what children observe in the local environment.

In discussing her pedagogical strategy, Meghan said that sometimes the class will collectively choose a single, burning question and she will model the inquiry process and work through it with the class. At other times, she will post all of the student questions—"all of the wonders they have had, and one wonder would jump to another wonder"—from which students choose a series of questions they are personally connected to and interested in exploring in more depth. Meghan may also post the series of questions for the class but then go onto something else, returning to discuss the questions in the following weeks. The specific process is guided by her experience and understanding of the curriculum, her students, and what possibilities the questions open for the class. Yet within these possibilities, Meghan is never certain what will emerge since she is often "learning alongside [her] students."

While there is no set approach that best corresponds to standard situations, Meghan relies on her practical wisdom as an educator—her *phronesis* (Gadamer, 2006) —to cultivate questioning and discovery in the most seamless, integrated way possible. Students' questions and wonderments became an essential part of their experience, as opposed to a distraction from the outcomes predetermined by the teacher (Jardine, 2012; Pelech, 2015). Meghan is refreshingly candid in describing her PCP as not set in stone and not fully formalized. Nevertheless, the educational value of continuing to teach within this uncertainty and embracing the unexpected moments of place has not gone unnoticed. Karen describes how Meghan openly shares with colleagues her concern for authentic learning while attending to the multitude of standard curricular outcomes, which has resulted in her becoming a key person who motivates other teachers to think about the authenticity of their place-conscious practice. Meghan's knowledge of the program of studies allows her to point out the multiple ways students are learning and is a source for rethinking and revising different aspects of place-conscious practices and projects.

Sometimes, from the wonderments, a much larger question will emerge that helps guide the learning for an extended period or even the whole year. Again, having a clear understanding of the curriculum and being open and flexible to students' voices are key characteristics of Meghan's work. As an example, Meghan described how, after being outside on one occasion and exploring what students wondered about, a larger question arose: What is the circle of life and how are the living things in their natural environment interconnected? This question guided their learning in science for the rest of the year. For part of the time, the class explored this question as a large group; at other times, students worked independently on one element of the question. For example, students initially created a large diorama that covered the back wall of the classroom. As students researched the variety of local animal species, they began to look more and more at how the animals were connected to the others. They began to ask questions about what should be next to the deer in order to demonstrate the connections between the scientific categories of producers, consumers, and decomposers. The diorama became an intricate representation of the interconnections between the living things in the area; an ecology and food web grounded in the specific place was the result. Meghan said "the students were able to begin to understand how complex the connections of all of the plants and animals in the area were." Compared to the generic food chain discussed earlier, students developed a richer understanding of the concept of food chains and webs and saw how these fit within the local ecology. This richer understanding of ecology stems from identifying and navigating the interconnections present in their environment. As students were drawn into paying more attention to their surroundings, they realized that life is far more complex than a one-dimensional, unidirectional chain of events. At the same time, students were cultivating an affinity for their surroundings and the realization that they, too, were part of the interconnections. This place-conscious practice began to reveal a "deep understanding of complex, diversity, and yet co-existing relationships to their habitat … fusing familiar non-formal cultural knowledge with scientific theory" (Avery & Hains, 2017, p. 131).

12.7   Connecting Students to the Local and Scientific Communities

The initial inquiry into the local ecosystem became a full-year project that included inviting scientists and other experts from the community to become integral participants in student learning. This building of school-community relationships is another key benefit of place-conscious pedagogy. Community engagement invites teaching and learning to come from a myriad of local sources including "community members, teachers, elders, parents, community experts, researcher, and youth in all aspects of the research" (Bang & Medin, 2010, p. 1024). This openness to community aligns with Alberta Education's curriculum redesign framework: "learning is embedded in relationship, culture, family, Elders, Knowledge Keepers, community, land, connections, memory and history" (Alberta Education, 2017, p. 5).[1] As a result, students are able to connect and deepen their community-based ways of knowing in conjunction with Western scientific understandings. These connections move the student experience from focusing solely on science content to offering opportunities as integral agents in designing and implementing their learning environments (Bang & Medin, 2010). As Karen expressed to Meghan, "One of the things I love that you do is how you bring in all the experts and all these people like bee experts, owl experts, to the class to help deepen the inquiry." Gold et al. (2015) reported that, when students are exposed to a variety of experts in a field of study such as ecology, then student engagement in the scientific process increases and student awareness of who does science expands to include women, visible minorities, knowledge keepers, and elders.

Along one pathway in the neighbourhood forest, a local community organization had posted pictures and information about the wildlife in the area. Meghan's students became curious about these posters and wanted to know more about the animals depicted. In response to their wonderings, Meghan contacted one of the people involved in producing the posters to ask if they would meet with her class. The students were excited to hear more firsthand information about local wildlife from the Community Recreation Coordinator; during their discussion, Meghan learned there was a $2,000 grant available for students to publish their work. Meghan and the class decided to apply for the grant, and the students set about learning the "who, what, where, when, and whys of applying for a grant. The students wrote and completed the different parts of the grant." This process connected their science studies to other subjects like mathematics as they began to form the idea of publishing and selling a book to the community about animals in the forest. The students worked with a local bank that helped them calculate expenses, revenues, and other important elements of publishing and selling. The project connected with the art curriculum when students drew large coloured pictures of the animals, which led them to a deeper understanding of biology and the science behind the study of anatomy and physiology. In this way,

---

[1] We have chosen to cite this version of the Alberta Education Program of Study documents because the 2022 draft was under critical scrutiny by educators, administrators and curriculum theorists at the time of writing.

the students experienced the cross-curricular dimension of their questions and saw how science is connected to other subject areas as well as to the community.

The students' place-conscious learning experience transformed into a community-wide learning engagement where the students became the experts. What started as a walk that sparked curiosity and questions resulted in a picture book of local animals that was published and sold within the school community (see Figs. 12.1 and 12.2). Students were proud and motivated to share their knowledge with a variety of people with the publication of their book. Meghan and the students organized a book launch in the school gymnasium; experts from a local nature centre provided taxidermy samples of the animal species the students had studied and included in their book so that they could display the animal they researched. Each student acted as a wildlife interpreter for the species they had researched. As Meghan remembers, "There were owls, birds, foxes all mounted. Half the gym was packed with students and the whole school came through and learned about the animals and they could buy the book."

The students had become the experts, sharing their knowledge with the rest of the school, parents, and even local and provincial politicians who attended the event. The resulting sense of student empowerment was described by Meghan when she shared how the school had a dress-as-your-favourite-book-character day. On the day, she noticed many of her students were not dressed up. She asked, "Who did you dress up as?" Their response was "I dressed up as myself. I am an author of our own book!" Meghan said this was one of her most memorable moments in teaching and she will share it with the students when it comes time for them to graduate.

Further opportunities for real-life learning emerged when the class decided to build bat/bird boxes with their families, teaching their parents and siblings about the importance of the natural world and their role in it. Meghan shared, "It was a once in a lifetime event, one of those magical times.... Students wanted to build bird/bat boxes with their parents and grandparents. We had a grandfather in a wheelchair hammering away—a multigenerational building project." Parents, caregivers, family, and community members embraced this place-conscious way of learning and enthu-siastically participated in improving the local ecosystem. As Karen explained during the interview, "It helps parents understand that the schoolwork means something—that teachers are not offering [science] out of a packet of worksheets and that student voices matter."

12.8 Student Agency and Connection to Place

Jardine et al. (2003) posed questions that get at the heart of enhancing student agency: "What does it mean to lead our children carefully and generously into such territories [of mathematics, or science, or poetry]? How can teachers ... help students under-stand that their work can be a real part of these places and can make a real difference?" (p. 8). We are convinced that the answer to such questions can be found in the work of teachers like Meghan. Throughout the many examples Meghan provided, a key theme that emerged was nurturing students' sense of agency. She described how

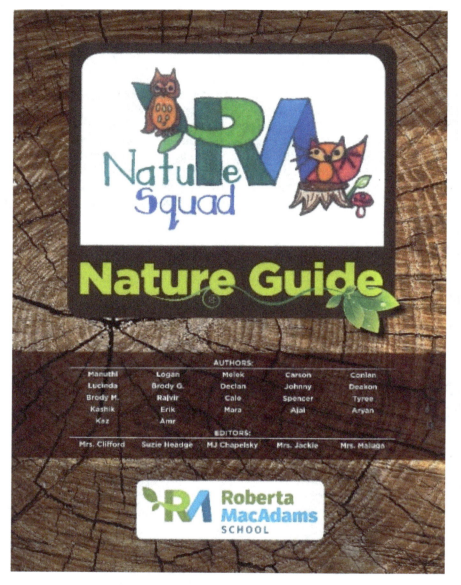

Fig. 12.1 Nature guide book published by Grade 4 students Roberta MacAdams School

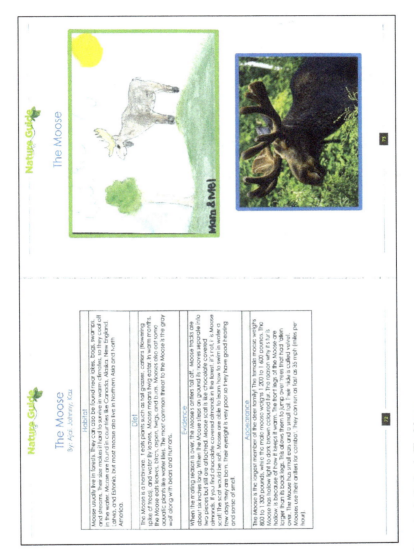

Fig. 12.2 A student entry in the book: The Moose

when she is with the kids, she is not "the knower; when in nature, "we sit in the circle on the grass—everyone is sitting in the circle, equally, and everyone has [an equal] voice." Meghan sees herself as a co-learner as the group asks questions together and every contribution is equally respected. "I have learned so much from kids and from being outside. I know some things, but then they will ask 'Why is that?' and I say, 'I don't know' (she laughs)." She guides the students through the process of how to learn and answer the questions they have posed. The students have direct involvement and agency not only in what they learn but also in how they go about learning. Research has shown that if students cannot see how science is personally relevant then they quickly become disengaged (Kalas & Raisinghani, 2019; Lyons, 2006; Page, 1999; Pelech, 2015). Conversely, when students are "actively creating, enacting and bringing forth meaning [as opposed to passively receiving knowledge, they become engaged and motivated to learn within an] anticipatory experience that engages the whole person" (Smits, 2003, p. 23), through an awakening of their innate curiosity.

Students who are given voice and a sense of agency for their learning can see how they can contribute and integrate with other members of a scientific community of inquiry. They learn that the work they do and the questions they ask are a key part of the scientific process—the same process scientists undertake in order to explore and understand the world around them. They can actively contribute to identifying and addressing local concerns. Simon and Osborne (2010) contended that what engages young people and activates their sense of agency is recognition of "their potential individual contribution(s) to the future" (p. 256).

Meghan also described how the students' understanding of their presence in the natural environment and self-awareness when in the forest had deepened beyond what she could have imagined. For example, the students learned how to conduct themselves while walking in a natural landscape. Meghan made clear:

> Students learn that if you want to hear the birds [and] all the other animals, you have to walk quietly. The beavers can feel our vibration when we are coming to see them so they will hide.... Some birds will do what is called a bird alarm and warn the other animals and birds. So when we saw the owl – the magpies have an incessant noise to warn the owl. [So students know] if we are not quiet, then we are not able to learn and see anything.

Hence, the students have raised their sensitivity and attention to place through interacting with nature and a desire to live and learn alongside the nonhuman members of their community. This awareness is taken home and shared with their families. In a letter to Karen, a kindergarten parent described the wonderful experience of entering nature with their child:

> I will try to mention a couple of things that REALLY warmed my heart ... the calmness to know how to stay quiet not to bother the birds and so they can gently become FRIENDS with the birds. [My child] showed me that approach when we went to explore the Porcupine Forest on a weekend, and I was blown away at her focus - so wonderful! We both lied [sic] there very quietly and sure thing - the birds came closer to us; she later told me all about how some birds had landed on her belly the other day and how cool that was.

From their affinity for the more-than-human world, the students learned they have an impact on the plants and animals of the area. From this, they developed an abiding respect for how they engage with place and an interest in changing their behaviour to fit within their new community, which is a clear connection to the science, technology, society, and environment decision-making outcomes in the curriculum.

The students' expanded sense of identity—their sense of who they are in the world—brought some very rewarding moments. Meghan shared that while on a walk with a biologist who was pointing out where great horned owls might be nesting, an owl flew right over the students. Swooping low, the owl landed across the river from where they stood. Jardine (2006) noted this ecological way of learning:

> Ecological awareness always and already involves the presence of our children. Ecology thus always already involves images of pedagogy and teaching and learning of the tales that need to be told for all of us to live well.… Understood in this Earthly, intergenerational way, education (and not just 'environmental education' as a subbranch, most often, of science education) has the opportunity, perhaps the obligation, to slow down the pace of attention, to broaden out its own work into the long-standing patterns and places we inhabit and that inhabit us. It has the opportunity, perhaps the obligation, to take on a mood not unlike ecological mindfulness. (p. 180)

By creating the opportunity for students to experience the unexpected, the arrival of the owl, Meghan embodied the pedagogical awareness needed for an ecological way of learning that Jardine describes. A teacher cannot plan what will happen, but instead can create the space for something to emerge through their ecological mindfulness. As Jardine argued, the students' knowledge of, affinity for, and relation to the place they inhabit goes well beyond that of using the environment for the purpose of science education. Instead, the students have placed the curriculum in their world and brought it to life through a reanimation of its meaning. Moreover, they are working daily to embody what it means to live well within a healthy ecology. King and Glackin (2010) contended that experiences of learning with nature and engaging with place offer authentic, first-hand opportunities that push beyond the traditional boundaries of school science. Students do not simply learn about local plants and animals; they know how to live well within these places; they cultivate a deeper understanding and agency that cannot be replicated in the classroom. Meghan's students are a hopeful example of agency for living well within the local ecology.

12.9   Covering the Curriculum and Ensuring Rigour

Whenever teaching and learning move away from the routine acts of schooling, questions of rigour and content arise. Inquiry and PCP are often accused of generating a free-for-all whereby students are asked to learn complex scientific phenomenon on their own or where there is no operative conception of curriculum standards (Jardine et al., 2003). From our research we have been shown time and again how engagement of students with a topic that emerges from place demands rigorous attention for students to treat their questions well. This case study shows that a key aspect of a place-conscious approach to teaching rests in developing authentic student interests into questions worthy of serious study—questions that researchers in the scientific community may be investigating. Engaging students through questions that emerge from place is often described as "student-centered learning" (Pelech, 2015 p. 213). However, Dewey (1902) warned that student-centered learning can end up becoming an empty pedagogy if undertaken without due regard of curriculum and process. He described the misconception that student-centered learning means students would be left alone to discover and make sense of their own world, which cannot be further from the truth. Dewey argued: "Development does not mean just getting something out of the mind. It is a *development of experience* [emphasis added] and into experience that is really wanted" (p. 18). Through the guided exploration of relevant, empirical questions, the space for authentic learning opens; in this space, students develop an experience of inquiry and a disposition for learning initiated by the experience and continued by engaging in the learning of scientific concepts. The teachers' expertise and the inclusion of knowledgeable community members help ensure the student's experience is embedded in authentic scientific understandings.

Meghan described how, during nature walks, she balances the openness and flexibility of encouraging the interests of the students with her knowledge of curriculum and expected outcomes. She said that she approaches learning through nature thoughtfully, "It is important to not just go on a random walk." Meghan is more than just a *guide on the side*, which is how teachers are often positioned in student-centered learning; she is a well-informed, active developer of student experience with a rich understanding of the curriculum, her students, and their community. From this situated expertise, she can create the conditions necessary for students to experience an authentic sense of exploration and discovery. This educative practice is not simply a method but a true pedagogy; it is a conscious way of teaching in close relation to curricular knowledge and the place where Meghan teaches (Pelech, 2015).

Karen shared how she was a little nervous at first about what parents might think of teaching in this way: Might parents see nature walks as frivolous and a waste of valuable school time? Karen's own pedagogy was already deeply rooted in PCP and authentic learning. Her concern, therefore, was not for her own practice but focused on educating parents so they might understand the high-quality education students were receiving. During the interview, she described the process of informing parents:

> I started with [educating] parents with the [curricular] competencies right away – I have been putting them out in the newsletter the first year, second year showing them what they look like by using examples from the classroom. Teachers were also doing that on their own [through] celebration of student learning centered on the competencies. This was refreshing that the parents embraced it.

Tytler (2010) reported that parents can often be conservative voices holding to more traditional conceptions of science education. He argues that it is imperative for teachers and administrators to work within educational systems to garner support from parents, trustees, and school boards; he recommends this be undertaken slowly through a process of systematically educating key stakeholders in the school system. Informing parents of the competencies, connecting learning with the curriculum, as well as being able to point to research that demonstrates the positive impacts of PCP have all helped gain parental support and confidence in the quality of their children's learning experiences. Karen noted the school rarely needs to defend its pedagogy to parents or other stakeholders. The high quality of their educational program is further seen in the increasing number of requests to attend the school from outside the catchment area, from none in the first year the school opened to over 30 in its fifth year.

12.10 Concluding Thoughts

From this and other cases of place-conscious pedagogy we have studied, five central themes continue to surface and inform our understanding of its value for science education. Within the myriad ways teachers take to place-conscious pedagogy, common themes are shared across settings in that it consistently enhances the transformative agency of teachers and their students; affirms the professional identity of teachers and the personal identity of students; supports diverse methods and understandings of teaching, learning, and assessing; strengthens teacher-student and school-community relationships; and provides a source of meaning, hope, and resilience.

For teachers who may not have access to forests, creeks, and other natural settings, Meghan offers encouragement for practicing PCP nearly anywhere:

> In our school field we have four trees; and by going to and observing those four trees over time (September to June), the questions that can come from looking and observing are fantastic! Another example is from going for a walk [when] a student found a feather on the pavement. This finding led to all types of questions from the type of bird to bird flight, aerodynamics, structure of feathers, etc. So these are great examples for teachers to see it can come from anywhere, even just sitting on the lawn; they don't need a forest; inspiration can come from anywhere.

Connecting with what students notice and wonder about is at the heart of place-conscious pedagogy. Whether in an urban or rural setting, place offers multiple ways to engage students with science. "[T]here are as many natural worlds and senses of place as there are different people. Place is a multitude of voices that tell places rather than a single voice" (van Eijck & Roth, 2010, p. 880).

When asked how teaching through place consciousness affected her sense of identity as a teacher, Meghan eagerly responded:

> One of the reasons I have gotten into teaching is I love seeing those lightbulb moments when a student is curious about something and when they discover something.... And when you are outside those moments are endless.... I have seen this grow as an incredible connection for me personally but also when I am with my students.... It is magical, it is just magical to watch the students learn.

Meghan's response echoes those of other teachers we interviewed. Their sense of identity as teachers is invested in the educational value of place. As Karen shared, "Place-conscious pedagogy helps reshape our world, and our acceptance and sense of belonging as a people." Place transforms students and educators as their awareness, affinity, and wonder grows. We hear the voice of enchantment in every teacher we interviewed and see the magic in every instance of place-conscious pedagogy. As Meghan described:

> I tell my students that I try to grow their brains over the year, but I also try and grow their hearts. I see a large part of that is taking them outside and seeing them understand that we all have a role and they are important in this role and so they too can understand the impact they can make, and they have made, and they continue to make with other people in order to help preserve place.

References

Addison Wesley. (2001). *Science in action 7*. Pearson Education Canada.

Aikenhead, G. S. (2003). *Review of research on humanistic perspectives in science curricula.* Paper presented at the European Science Education Research Association, Noordwijkerhout, The Netherlands. https://education.usask.ca/documents/profiles/aikenhead/ESERA_2.pdf

Alberta Education. (1996). *Program of studies science (K–6)*. https://education.alberta.ca/media/159711/elemsci.pdf

Alberta Education. (2017). *The guiding framework for the design and development of kindergarten to grade 12 provincial curriculum.*

Asfeldt, M., Urberg, I., & Henderson, B. (2009). Wolves, ptarmigan, and lake trout: Critical elements of a Northern Canadian place-conscious pedagogy. *Canadian Journal of Environmental Education, 14*, 33–41.

Avery, L. M., & Hains, B. J. (2017). Oral traditions: A contextual framework for complex science concepts–laying the foundation for a paradigm of promise in rural science education. *Cultural Studies of Science Education, 12*(1), 129–166. https://doi.org/10.1007/s11422-016-9761-5

Babbie, E., & Benaquisto, L. (2010). *Fundamentals of social research* (2nd Canadian ed.). Nelson.

Bang, M. E., & Medin, D. (2010). Cultural processes in science education: Supporting the navigation of multiple epistemologies. *Science Education, 94*(6), 1008–1026. https://doi.org/10.1002/sce.20392

Basu, S. J., & Barton, A. C. (2010). A researcher-student-teacher model for democratic science pedagogy: Connections to community, shared authority, and critical science agency. *Enquiry & Excellence in Education, 43*(1), 72–87. https://doi.org/10.1080/10665680903489379

Biesta, G., & Tedder, M. (2007). Agency and learning in the lifecourse: Towards an ecological perspective. *Studies in the Education of Adults, 39*(20), 132–149.

Bigelow, B. (2014). How my schooling taught me contempt for the Earth. In B. Bigelow & T. Swinehart (Eds.), *A people's curriculum for the Earth: Teaching climate change and the environmental crisis* (pp. 36–41). Rethinking Schools.

Blades, D. W. (1997). *Procedures of power and curriculum change: Foucault and the quest for possibilities in science education*. Peter Lang.

Blades, D. W. (2001). The simulacra of science education. In J. A. Weaver, M. Morris, & P. Appelbaum (Eds.), *(Post)Modern science (education): Propositions and alternative paths* (pp. 57–94). Peter Lang.

Chang, D. (2017). Diminishing footprints: Exploring the local and global challenges to place-based environmental education. *Environmental Education Research, 23*(5), 722–732.

Collyer, C. (1998). *All hands in the dirt: A guide to creating natural school grounds*. Evergreen Foundation.

Coughlin, C. A., & Kirch, S. A. (2010). Place based education: A transformative activist stance. *Cultural Studies of Science Education, 5*(4), 911–921. https://doi.org/10.1007/s11422-010-9290-6

Crease, R. P. (1997). Hermeneutics and the natural sciences. *Man and World, 30*(3), 259–270. https://doi.org/10.1023/A:1004221117249

Creswell, J. W., & Miller, D. M. (2000). Determining validity in qualitative inquiry. *Theory into Practice, 39*(3), 124–130. https://doi.org/10.1207/s15430421tip3903_2

Deringer, S. A. (2017). Mindful place-based education: Mapping the literature. *Journal of Experiential Education, 40*(4), 333–348. https://doi.org/10.1177/1053825917716694

Dewey, J. (1902). *The child and the curriculum.* University of Chicago Press.

Eger, M. (1993). Hermeneutics as an approach to science: Part 1. *Science & Education, 2,* 1–29.

Gadamer, H.G. (2006). *Truth and method* (2nd rev. ed.) (J. Wiensheimer & D.G. Marshall, Trans.). Continuum.

Gold, A. U., Oonk, D. J., Smith, L., Boykoff, M. T., Osnes, B., & Sullivan, S. B. (2015). Lens on climate change: Making climate meaningful through student-produced videos. *Journal of Geography, 114*(6), 235–246. https://doi.org/10.1080/00221341.2015.1013974

Goodman, J. F., & Eren, N. S. (2013). Student agency: Success, failure, and lessons learned. *Ethics and Education, 8*(2), 123–139. https://doi.org/10.1080/17449642.2013.843360

Greenwood, D. A. (2013). A critical theory of place-conscious education. In R. B. Stevenson, M. Brody, J. Dillon, & A. E. J. Wals (Eds.), *International handbook of research on environmental education* (pp. 93–100). Routledge. https://doi.org/10.4324/9780203813331

Gruenewald, D. A. (2003). Foundations of place: A multidisciplinary framework for place-conscious education. *American Educational Research Journal, 40*(3), 619–654. https://doi.org/10.3102/000 28312040003619

Gruenewald, D. A. (2008). The best of both worlds: A critical pedagogy of place. *Environmental Education Research, 14*(3), 308–324. https://doi.org/10.1080/13504620802193572

Gruenewald, D. A., & Smith, G. A. (eds.). (2010). *Place-based education in the global age.* Routledge. https://doi.org/10.4324/9781315769844

Harvard Graduate School of Education. (1999a). *Living and learning in rural schools and communities: A report to the Annenburg Rural Challenge.*

Harvard Graduate School of Education. (1999b). *Living and learning in rural schools and communities: Lessons from the field.*

Jardine, D. W. (2006). On hermeneutics: 'Over and above our wanting and doing.' In K. Tobin & J. L. Kincheloe (Eds.), *Doing educational research: A handbook* (pp. 269–288). Sense.

Jardine, D. W. (2012). *Pedagogy left in peace: Cultivating free spaces in teaching and learning.* Continuum.

Jardine, D. W., Clifford, P., & Friesen, S. (2003). *Back to the basics of teaching and learning: Thinking the world together.* Lawrence Erlbaum Associates.

Kalas, P., & Raisinghani, L. (2019). Assessing the impact of community-based experiential learning: The case of Biology 1000 Students. *International Journal of Teaching and Learning in Higher Education, 31*(2), 261–273. https://www.isetl.org/ijtlhe/pdf/IJTLHE31(2).pdf

Kelly, D. (2005). On reconceiving our school grounds. *Pathways: The Ontario Journal of Outdoor Education, 18*(1), 31–34. https://www.coeo.org/wp-content/uploads/pdfs/Digital_Pathways/Pat hways_18_1.pdf

Kelly, D. (2006). An ecological conception of school-ground education. *Interactions: The Ontario Journal of Environmental Education, 19*(1), 28–29.

Kelly, D. (2007). Thoughts on content and method in place-based education. *Interactions: The Ontario Journal of Environmental Education, 19*(4), 9–12.

Kelly, D., & Pelech, S. (2019a). A critical conceptualization of place-conscious pedagogy. *European Journal of Curriculum Studies, 5*(1), 732–741. http://pages.ie.uminho.pt/ejcs/index.php/ejcs/art icle/view/187/100

Kelly, D., & Pelech, S. (2019b). From port and prairie: Exploring the impact of place-conscious pedagogy on student agency. *European Journal of Curriculum Studies, 5*(2), 874–885. http:// pages.ie.uminho.pt/ejcs/index.php/ejcs/article/view/202/110

King, H., & Glackin, M. (2010*)*. Supporting science learning in out-of-school contexts. In J. Osborne & J. Dillon (Eds.), *Good practice in science teaching: What research has to say* (2nd ed., pp. 259–273). McGraw-Hill.

Kudryavtsev, A., Steman, R. C., & Krasney, M. E. (2012). Sense of place in environmental education. *Environmental Education Research, 18*(2), 229–250. https://doi.org/10.1080/13504622.2011.609615

Lescure, K., & Yaman, C. (2014). Place-based education: Bringing schools and communities together. In C. White (Ed.), *Community education for social justice* (pp. 23–31). Sense.

Liberman, J., & Hoody, L. (1998). Closing the achievement gap: Using the environment as an integrating context for learning. *State Education and Environmental Roundtable.*

Lynch, J., Eilam, E., Fluker, M., & Augar, N. (2017). Community-based environmental monitoring goes to school: Translations, detours and escapes. *Environmental Education Research, 23*(5), 708–721. https://doi.org/10.1080/13504622.2016.1182626

Lyons, T. (2006). Different countries, same science classes: Students' experiences of school science in their own words. *International Journal of Science Education, 28*(6), 591–613.

Nichols, J. B., Howson, P. H., Mulrey, B. C., Ackerman, A., & Gately, S. E. (2016). Promise of place: Using place-based education principles to enhance learning. *International Journal of Pedagogy and Curriculum, 23*(2), 27–41. https://doi.org/10.18848/2327-7963/CGP/v23i02/27-41

Ormond, C. G. A. (2013). Place-based education in practice. In D. B. Zandvliet (Ed.), *The ecology of school: Advances in learning environments research* (pp. 19–28). Sense.

Page, R. N. (1999). The uncertain value of school knowledge: Biology at Westridge High. *Teachers College Record, 100*(3), 554–601. https://www.tcrecord.org. ID Number: 10323.

Pelech, S. (2015). *What does it mean to teach biology well? A hermeneutic inquiry.* Unpublished doctoral dissertation, University of Calgary.

Pelech, S., & Kelly, D. (2020). Teacher identity and agency: Learning and becoming through place-conscious pedagogy. In E. Lyle (Ed.), *Identity landscapes: Contemplating place and the construction of self* (pp. 210–221). Brill/Sense.

Pelo, A. (2014). A place for ecology. In B. Bigelow & T. Swinehart (Eds.), *A people's curriculum for the Earth: Teaching climate change and the environmental crisis* (pp. 42–47). Rethinking Schools.

Piersol, L. (2013). Local wonders. In D. B. Zandvliet (Ed.), *The ecology of school: Advances in learning environments research* (pp. 63–72). Sense.

Pirrie, A., & Gillies, D. (2012). Untimely meditations on the disciplines of education. *British Journal of Educational Studies, 60*(4), 387–402. https://doi.org/10.1080/00071005.2012.727380

Schensul, S. L., Schensul, J. J., & LeCompte, M. D. (1999). *Essential ethnographic methods: Observations, interviews, and questionnaires.* Alta Mira Press.

Sedawi, W., Assaraf, O. B. Z., & Reiss, M. J. (2021). Regenerating our place: Fostering a sense of place through rehabilitation and place-based education. *Research in Science Education, 51*(Suppl. 1), 461–498.

Seidman, I. (2019). *Interviewing as qualitative research: A guide for researchers in education and the social sciences* (5th ed.). Teachers College Press.

Semken, S., & Butler Freeman, C. (2008). Sense of place in the practice and assessment of place-based science teaching. *Science Education, 92*(6), 1042–1057. https://doi.org/10.1002/sce.20279

Semken, S., & García, Á. A. (2021). Synergizing standards-based and place-based science education. *Cultural Studies of Science Education, 16*(2), 447–460.

Simon, S., & Osborne, J. (2010). Students' attitudes to science. In J. Osborne & J. Dillon (Eds.), *Good practice in science teaching: What research has to say* (2nd ed., pp. 238–258). McGraw-Hill Press.

Skilbeck, A., & Stickney, J. (2020). Section 5. *Journal of Philosophy of Education, 54*(4), 1032. https://doi.org/10.1111/1467-9752.12518

Smith, G. A. (2002). Place-based education: Learning to be where we are. *Phi Delta Kappan, 83*(8), 584–594. https://doi.org/10.1177/003172170208300806

Smith, G. A. (2007). Place-based education: Breaking through the constraining regularities of public school. *Environmental Education Research, 13*(2), 189–207. https://doi.org/10.1080/135046207 01285180

Smits, H. (2003). *Trying to teach, trying to learn: Listening to students.* Alberta Teachers' Association. https://www.teachers.ab.ca/SiteCollectionDocuments/ATA/Publications/Benefits-and-Wor king-Conditions/PD-14a%202008%2008%2027.pdf

Sobel, D. (1996). *Beyond ecophobia: Reclaiming the heart in nature education.* Orion Society & Myrin Institute.

Sobel, D. (2005). *Place-based education: Connecting classrooms & communities.* Orion Society.

Sobel, D. (2008). *Childhood and nature: Design principles for educators.* Stenhouse.

Tytler, R. (2010). Stories of reform in science education: Commentary on opp(reg)ressive policies and tempered radicals. *Cultural Studies of Science Education, 5*(4), 967–976. https://doi.org/10. 1007/s11422-010-9277-3

University of Alberta. (2021). *Acknowledgement of traditional territory.* https://www.ualberta.ca/ toolkit/communications/acknowledgment-of-traditional-territory

Üztemur, S., & Dere, İ. (2022, advanced online publication). 'I was not aware that I did not know': Developing a sense of place with place-based education. *Innovation*, 1–17. https://doi.org/10. 1080/13511610.2022.2092457

van Eijck, M., & Roth, W.-M. (2010). Towards a chronotopic theory of "place" in place-based education. *Cultural Studies of Science Education, 5*(4), 869–898. https://doi.org/10.1007/s11 422-010-9278-2

Wong, S. L., & Hodson, D. (2010). More from the horse's mouth: What scientists say about science as a social practice. *International Journal of Science Education, 32*(11), 1431–1463. https://doi. org/10.1080/09500690903104465

Sharon Pelech BEd (University of Alberta), MEd (University of Lethbridge), PhD (University of Calgary) is an associate professor of science education in the Faculty of Education at the University of Lethbridge. She teaches at the undergraduate and graduate levels in curriculum studies and science education. Previous to this, Sharon was an assistant professor at Memorial University of Newfoundland and was an instructor at the University of Calgary. Sharon spent over 18 years as a secondary science teacher in the Northwest Territories, Northern Alberta, and Calgary. Her experiences as an educator in a variety of locations has influenced her research interests in science education, ecopedagogy, place-conscious pedagogy and interpretive (hermeneutic) research.

Darron Kelly BA (Mount Allison), BEd (Queen's), MEd (Queen's), PhD (Memorial) is an associate professor with Memorial University, Faculty of Education. Darron is an award-winning, SSHRC-funded scholar who examines applications of critical social theory in educational administration (including communicative rationality, transformative leadership, and moral policy-making) and explores the educational value of place-conscious pedagogy (including teacher identity, school-community partnerships, and transformative student agency).

Index

CPSIA information can be obtained
at www.ICGtesting.com
Printed in the USA
LVHW082309050423
743641LV00002B/26